Operation of Extended Aeration Package Plants

WEF Manual of Practice No. OM-7

Second Edition

Prepared by **Operation of Extended Aeration Package Plants**
Task Force of the **Water Environment Federation**

Keith Radick, *Chair*

John Brown	Paul Eckley	Allan Townshend
William Busch	J.C. Goldman	Jeff Vaughn
William Callegari	Charlie Liu	Wesley Warren
John Chalas	Jerry McClary	Richard Weigand
James Clifton	Lewis N. Powell	
Ernest Earn	Keith Radick	

T0155292

Under the Direction of the **Municipal Subcommittee** of the **Technical Practice Committee**

2004

Water Environment Federation
601 Wythe Street
Alexandria, VA 22314–1994 USA
www.wef.org

IMPORTANT NOTICE

The material presented in this publication has been prepared in accordance with generally recognized engineering principles and practices and is for general information only. This information should not be used without first securing competent advice with respect to its suitability for any general or specific application.

The contents of this publication are not intended to be a standard of the Water Environment Federation (WEF) and are not intended for use as a reference in purchase specifications, contracts, regulations, statutes, or any other legal document.

No reference made in this publication to any specific method, product, process, or service constitutes or implies an endorsement, recommendation, or warranty thereof by WEF.

WEF makes no representation or warranty of any kind, whether expressed or implied, concerning the accuracy, product, or process discussed in this publication and assumes no liability.

Anyone using this information assumes all liability arising from such use, including but not limited to infringement of any patent or patents.

Library of Congress Cataloging-in-Publication Data

Operation of extended aeration package plants / prepared by Operation of Extended Aeration Package Plants Task Force of the Water Environment Federation.—2nd ed.
    p. cm.—(WEF manual of practice; no. OM-7)
  Rev. ed. of: Operation of extended aeration package treatment plants.
  Includes bibliographical references and index.
  ISBN 978-1-57278-191-7
  1. Aerated package treatment systems—Handbooks, manuals, etc. I. Water Environment Federation. Operation of Extended Aeration Package Plants Task Force. II. Operation of extended aeration package treatment plants. III. Manual of practice. Operations and maintenance; no. OM-7.
  TD779.O64 2004
  628.3′54—dc22   2004013587

Water Environment Federation

Improving Water Quality for 75 Years

Founded in 1928, the Water Environment Federation (WEF) is a not-for-profit technical and educational organization with members from varied disciplines who work toward the WEF vision of preservation and enhancement of the global water environment. The WEF network includes water quality professionals from 79 Member Associations in over 30 countries.

For information on membership, publications, and conferences, contact

Water Environment Federation
601 Wythe Street
Alexandria, VA 22314-1994 USA
(703) 684-2400
http://www.wef.org

Manuals of Practice
of the Water Environment Federation

The WEF Technical Practice Committee (formerly the Committee on Sewage and Industrial Wastes Practice of the Federation of Sewage and Industrial Wastes Associations) was created by the Federation Board of Control on October 11, 1941. The primary function of the Committee is to originate and produce, through appropriate subcommittees, special publications dealing with technical aspects of the broad interests of the Federation. These publications are intended to provide background information through a review of technical practices and detailed procedures that research and experience have shown to be functional and practical.

Water Environment Federation Technical Practice Committee
Control Group

G.T. Daigger, *Chair*
B.G. Jones, *Vice-Chair*

R. Karasiewicz
M.D. Nelson
A.B. Pincince

Authorized for Publication by the Board of Directors
Water Environment Federation

William Bertera, *Executive Director*

Contents

List of Tables

List of Figures

Preface

The treatment processes described in this manual are typical of small package extended aeration wastewater treatment plants, many of which are prefabricated and shipped to the construction site as one unit. The extended aeration process is a variation of the activated sludge treatment process. The activated sludge process is a two-stage system consisting of aeration tanks and secondary clarifiers. The aeration tank receives primary effluent or raw wastewater and provides it with a controlled environment in which organic wastes are consumed by microorganisms in an efficient manner. The secondary clarifier (also referred to as a *settling tank* or *sedimentation tank*) facilitates the separation of the suspended solids from the treated wastewater. The clear liquid at the top of the settling chamber will then typically flow to another chamber for disinfection. This manual is intended for the individual who is operating a treatment plant with a design flow of up to 5 L/s (100 000 gpd), typically without a laboratory or operations building on site. Chapter 10, Advanced Unit Processes, was not updated for this edition.

This second edition of this manual was produced under the direction of Keith Radick, *Chair*.

Principal authors of the publication are

William Callegari	(7)
John Chalas	(8)
Ernest Earn	(12)
Paul Eckley	(3, 11)
Mike Luker	(13)
Jerry McClary	(6)
Lewis N. Powell	(1, 2)
Keith Radick	(5)
Wesley Warren	(9)
Richard Weigand	(4)

Authors' and reviewers' efforts were supported by the following organizations:

Blue Heron Environmental Technology, Athens, Ontario
City of Salem, Oregon
Environmental Training Center, Ripley, West Virginia
Floyd Browne Associates, Inc., Marion, Ohio
GRW Engineers, Lexington, Kentucky
Georgia Department of Natural Resources, Lawrenceville, Georgia
Hampton Roads Sanitation District, Virginia Beach, Virginia
Metcalf & Eddy, Wakefield, Massachusetts
Simsbury WPC, Simsbury, Connecticut
State Department of Health, Charleston, West Virginia
United Water, Huntsville, Alabama
Vaughn, Coast & Vaughn, Inc., Saint Clairsville, Ohio

Chapter 1
Introduction

As our nation grew and became industrialized, areas of growth caused concentrations of businesses and populations. These concentrations produced quantities of industrial and human wastes that pit privies and septic tanks could no longer treat adequately. This led to the development of sewers as a means to convey wastes away from population centers.

Early systems, called *combined sewers*, collected both wastes and runoff from rainfall and melting snow and ice. Later, systems were developed to carry either waste or surface runoff and were called *sanitary sewers or storm sewers*, respectively. The type of collection system that discharges to the wastewater treatment plant is an important factor and will be discussed later in this manual.

Wastewater collection systems eliminated many health problems associated with urban life; however, all the wastes were being discharged from a few pipes to the nearest stream. As areas developed, more waste was generated and new towns developed along rivers, thereby reducing the distances between discharges. Eventually, the streams could no longer handle the waste without showing damaging effects. Rivers began to have foul odors and were often covered with floating solids. In some cases, the pollution caused the death of fish and other aquatic life.

Wastewater treatment processes were developed by observing what happened in nature. As waste entered the stream, the dissolved oxygen in the water decreased and bacteria populations increased. As the waste moved downstream, the bacteria eventually consumed all the organic material. Bacterial populations then decreased, the dissolved oxygen in the stream was replenished, and the whole process was repeated at the next wastewater discharge point. This natural process worked well until the distance between discharge points became so close that the dissolved oxygen content in the river could not be replaced quickly enough through natural means.

Waste material in domestic wastewater is generally organic (biodegradable), which means that microorganisms can use this matter as their food source. As aerobic microorganisms consume the organic material, oxygen is also consumed. An indication of the amount of organic material that can be consumed by the aerobic microorganisms is called the *biochemical oxygen demand* (BOD). By measuring the amount of oxygen the bacteria consume, the amount of biodegradable material in the waste may be determined.

Biological wastewater treatment processes work similarly to processes occurring in nature. Figure 1.1 shows the major units of a typical package extended aeration wastewater treatment plant. Wastewater is first screened to remove large floating, suspended, or settleable material that would interfere with the operation of the wastewater treatment plant. Materials typically removed by screening include leaves, sticks, rags, rubber, plastics, and rocks. After screening, wastewater may pass through a grinder to reduce large particles to small ones. This is done because microorganisms will digest smaller particles more readily than larger ones. As mentioned before, oxygen is needed when aerobic microorganisms consume organic material; therefore, a chamber is necessary to mix microorganisms with wastewater and provide oxygen to the bacteria. This chamber is referred to as the aeration chamber; it is some-

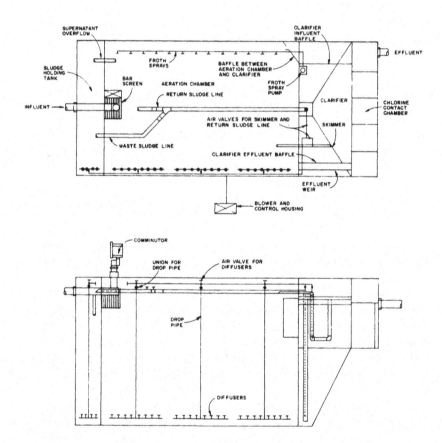

Figure 1.1   Basic units of a package extended aeration wastewater treatment plant.

times thought of as the living part of a wastewater treatment plant because it is here that the microorganisms multiply and grow. The liquid in this chamber, called mixed liquor, will have the consistency of a thin milkshake and a brown color similar to that of coffee with cream.

Because the goal of any good operating wastewater treatment plant is to discharge a clean, clear, safe liquid to the environment, the contents of the aeration chamber cannot be discharged without further treatment. If the liquid in the aeration chamber remains undisturbed in a container, most of the mixed liquor solids will settle to the bottom, leaving a clear liquid. The clear liquid on top (supernatant) will still contain numerous organisms (too small to settle) and dissolved solids. A settling chamber (clarifier) is placed after the aeration chamber to allow the microorganisms that are grown in the aeration chamber to settle by gravity. Most microorganisms settle to the bottom in the settling chamber and are then pumped back to the head end of the aeration chamber, where they again begin the cycle of feeding on incoming organics in the wastewater. This material is known as *return activated sludge*. The clear liquid at the top of the settling chamber will then typically flow to another chamber for disinfection.

Disinfection is the process of killing the disease-causing microorganisms (pathogens) that remain in the clear liquid. A disinfection unit is required for most wastewater treatment plants. Numerous disinfection agents, discussed later in the manual, may be used to kill pathogens.

After the wastewater has been through the treatment processes described, approximately 85 to 95% of the organic (biodegradable) material will have been removed. Regulatory agencies may require still further reduction of BOD before the treated wastewater can be discharged to the stream. Advanced (or tertiary) treatment typically means that the discharge from the wastewater treatment plant flows through sand filters, polishing ponds, or microstrainers.

The treatment process described in this chapter is typical of small package extended aeration wastewater treatment plants, many of which are prefabricated and shipped to the construction site as one unit. This type of construction is appropriately called *package* because the treatment plant is assembled complete and ready for shipping. Although some construction procedures allow wastewater treatment plants as large as approximately 45 L/s (1.0 million gpd) to be built using prefabricated units, only systems with design flows of approximately 5 L/s (100 000 gpd) or less will be discussed in this manual. Most small wastewater treatment plants are built by subdividing one large tank into an aeration chamber, clarifier, and disinfection unit.

Chapter 2
Installation

Package extended aeration wastewater treatment plants are typically constructed of concrete or steel. The systems are preengineered and preconstructed to minimize work at the installation site. If the system is small enough, the entire wastewater treatment plant will arrive at the site as a unit, ready to be set into the ground. For larger systems, the aeration chamber, clarifier, and chlorine contact tank will be delivered as separate units, and then assembled on site. Concrete plants may arrive as sections to be assembled on site, or may be constructed on site with preengineered panels.

A few basic rules should be followed during construction to ensure proper operation of all unit processes. Close inspection by the owner of the operator is essential.

The package plant is installed to accept wastewater flow from either an existing or newly completed collection system. The performance of the package plant will be only as good as its collection system. *Wastewater Collection Systems Management* (WEF, 1999) is recommended for additional collection system information.

Availability of a stream to receive discharge from the wastewater treatment plant and degree of treatment required by the regulatory agencies are important site location factors. The site should allow for wastewater collection by gravity flow where possible. The site should be accessible by vehicle year-round, well-drained, stable, and not subject to flooding. It should be served by potable (safe for drinking) water for hand washing, plant washdown, chemical mixing, and analytical testing. All hose spigots should be equipped with vacuum breakers to prevent accidental back-siphonage of wastewater into the potable water system. The plant should be at least 30 m (100 ft) from any residential area. In addition, the site may be governed by zoning ordinances or health department regulations.

From an operation and maintenance standpoint, a plant should be installed with the tank walls extending at least 150 mm (6 in.) above the ground. Such an installation allows ease of servicing, prevents surface runoff from entering the plant, and provides some insulation in colder weather. Plants installed completely below grade must have extension walls or well-defined diversion ditching to prevent surface runoff from entering the plant. Plants located completely above ground should have walkways provided around the entire plant. Above-ground plants will require insulation in cold climates.

A chain-link fence, 2.5 m (8 ft) in height, flush to the ground, with a locked entrance gate, and topped with three strands of angularly installed barbed wire, is recommended to prevent vandalism and tampering and discourage children from entering. A minimum of 1.2 m (4 ft) of working space between the fence and plant at all points is necessary for maintenance. A building constructed over the plant can be used instead of a fence if camouflage is desired or the plant is located in a cold climate. The building walls, like the fence, must be at least 1.2 m (4 ft) from any part of the plant to allow maintenance; the building must be ventilated, well-lit, and have adequate headroom above the plant (typically 2 m [7 ft]). The building will provide additional insulation and prevents pine needles, leaves, and other airborne debris from entering the plant. Housing also muffles noise from the blowers.

The plant should not be started until the manufacturer's and engineer's operation and maintenance manuals have been read and understood. The manufacturer should be required to provide training as part of the bid document. Training is recommended for all installations. Additional technical assistance should be provided by the design engineer.

Metal and concrete are the two most common materials used in the manufacture of package plants. Among the advantages of concrete are that it is less expensive, will not corrode if a short shipping distance is involved, and has a longer life expectancy. Some advantages of metal are that it is less subject to leakage, easier to service and alter, has a high salvage value and lighter weight, and there is less possibility of solids deposition between clarifier hopper and outer tank than might occur in concrete tanks because of poor field filleting.

Motors, blowers, and control panels should not be mounted on top of the plant. This is because on-plant locations interfere with the operator's maintenance procedures, all control equipment is exposed to a more corrosive atmosphere, and tools and equipment are more likely to be dropped into the plant while making repairs or servicing equipment.

The package wastewater treatment plant must be installed in a level position to work properly. A steel plant is typically placed on a level concrete pad that extends horizontally at least 0.3 m (1 ft) in all directions beyond the base of the plant. A concrete plant is typically set on damp, well-compacted, level sand. In all cases, the possibility of settling must be avoided. Because of the segmented design of concrete plants, it is important to eliminate any settling. A concrete plant requires the same foundation considerations as a steel plant.

During start-up or operation of the package plant, it may be necessary to dewater the plant to remove nonbiodegradable material such as dirt, sand, mud, or fibrous materials that may have entered the plant. Dewatering during construction should be done only after all sewers and lift stations have been thoroughly flushed to remove all material from the collection system. Dewatering should proceed from the influent end of the plant to the effluent end. Until the liquid level is below the opening between the aeration chamber and clarifier, the water level in both compartments will drop simultaneously.

Cleaning is not complete until all material is removed from all parts of the plant. Septic tank haulers may be used to dewater the plant. However, it may be necessary to remove mud and sand with shovels and buckets, because even relatively small amounts of mud or sand left in the plant can interfere with its operation. The treatment plant should be immediately refilled with water to prevent flotation after cleaning.

Because a package plant may need to be dewatered occasionally, care must be taken to prevent it from floating out of the ground from the buoyant force of water in the surrounding soil. Figure 2.1 illustrates how a steel plant is positioned and anchored to its foundation slab. Concrete plants are not typically anchored because of their weight; however, concrete plants can float and be damaged by hydrostatic uplift pressure. During construction, it is recommended that areas adjacent to all plants be backfilled with free-draining material (sand or gravel) and, where the lay of the land permits, a drain should

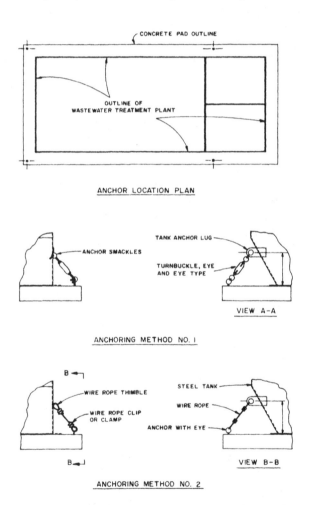

Figure 2.1   Typical anchoring methods of a steel plant to its foundation.

be installed to remove surrounding groundwater. If topography does not permit the installation of a drain, then well points should be placed around the plant so that the surrounding groundwater may be lowered before dewatering.

Well points can be constructed by installing 100- or 150-mm (4- or 6-in.) diameter plastic pipes, placed vertically at several locations around the plant, with a minimum of one pipe at each corner. The bottom 1.2 m (4 ft) of the pipe must be perforated. The top of the pipe should extend just above the ground surface and be capped loosely. The bottom of the pipe should be below the bottom of the plant. Gravel that is slightly larger than the holes in the pipe must be packed around it to allow water in and to keep the surrounding soil out of the pipe. It is easier to install the well points at the time the plant is being installed. The main objective is to provide a path for water to travel from the area around the plant to the well points.

If possible, a package plant should not be dewatered during wet weather. **An operating package plant cannot be dewatered before permission is obtained from the proper regulatory agencies.** It is imperative that drain plugs are installed on metal tanks before being set into the ground. Manufacturers leave the drain plugs out during fabrication and storage so that rainwater does not collect in the plant.

During final inspection, and while the plant is empty, all parts of the plant that are usually under water should be carefully inspected. Metal plants are typically painted with a coal tar epoxy paint to prevent corrosion. Any time an existing plant is dewatered, all surfaces that are usually submerged should be inspected for corrosion. If corrosion problems are corrected during the cleaning operation, dewatering at a future date to inspect for corrosion can be eliminated. It is also recommended that the design engineer inspect the plant at the fabrication site to eliminate field corrections.

The outside of all steel plants must be painted with coal tar epoxy or other suitable paint as recommended by the manufacturer. To minimize corrosion from acid soil, sacrificial magnesium anodes are buried in the ground approximately 3 m (10 ft) from the metal plant and attached to the base metal of the plant with a copper wire. Details on how the sacrificial anode works will be addressed in Chapter 11. During construction, proper installation of the anodes is critical. Not allowing enough slack in the copper wire, so that it breaks when the backfill around the plant settles, is a common mistake. Plenty of slack should always be left in the wire. Figure 2.2 shows proper installation of sacrificial anodes.

A frost-proof potable water hydrant with a backflow preventer should be located near the wastewater treatment plant for washdown and cleanup. On larger plants, clear water from the chlorine contact tank or clarifier may be used as washdown water; however, smaller plants do not have sufficient volume. Removal of supernatant from the clarifier of a small package plant will interfere with settling. Supernatant from the clarifier is used for foam control, which is discussed in Chapter 5.

Before testing and start-up, the owner should receive the manufacturer's operation and maintenance manuals for the plant, including all mechanical equipment. All appropriate manuals should be received and approved before

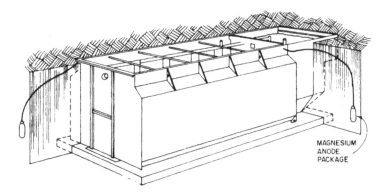

Note: Cathodic Protection. Magnesium anode packages are supplied with all prefabricated steel wastewater treatment plants for cathodic protection. These packages are to be securely connected to the aeration chamber. This is done by bolting the long heavy copper wire coming from the inside of the packaged anode to the connection lugs on the aeration chamber. This must be a good electrical connection. These packages are to be buried as deep and as far from the plant as possible while leaving plenty of slack in the wire. This is normally done when the plant is backfilled.

Figure 2.2   Installation of sacrificial anodes.

making final payment. Information in the manuals should include, but not be limited to, the following:

- A complete list of all replacement parts, including name of manufacturer and parts catalog number;
- Operating instructions;
- Maintenance instructions;
- Wiring diagrams;
- Troubleshooting procedures;
- As-constructed drawings; and
- Drawings of plan views showing the location of the treatment plant with respect to nearby permanent landmarks (buildings), all underground potable water and electric lines, underground influent and effluent lines, and sacrificial anodes.

REFERENCE

Water Environment Federation (1999) *Wastewater Collection Systems Management,* 5th ed.; Manual of Practice No. 7; Water Environment Federation: Alexandria, Virginia.

Chapter 3
Flow Equalization

GENERAL PURPOSE OF FLOW EQUALIZATION

All activated sludge wastewater treatment plants, including package extended aeration plants, depend on microorganisms (bacteria) to treat wastewater. Bacteria must have sufficient contact time with wastewater to reduce pollutants in wastewater to an acceptable level. For package extended aeration wastewater treatment plants, the design contact time between bacteria and wastewater is 18 to 24 h. This contact time and the daily flow determine the size of the aeration chamber, where oxygen (air) is supplied to the mixture of microorganisms and organic matter in the wastewater.

A package extended aeration plant is sized to accept the volume of wastewater produced during a 24-h period, measured in liters per second (or gallons per day). A constant flow to the plant during this period is ideal. The ideal or average flowrate for the 24-h period may be found by dividing the daily flow by 1440 min (1 d = 24 h = 1440 min). Unfortunately, wastewater flows are not constant, but vary during the day. Flowrates can also vary during the week.

Water use is highest in the morning (as people prepare for work or school) and during evening hours. Because a wastewater treatment plant is designed on the 24-h flow basis, it is critical that high (peak) flows do not cause sludge to be washed from the plant. If solids are lost during peak flow periods, then flow equalization may be required. Flow equalization is a method of temporarily storing wastewater during periods of excessive flows, then feeding it to the treatment plant uniformly throughout the 24-h period. Many regulatory agencies require flow equalization for treatment plants serving schools and campgrounds; however, flow equalization will probably not be required for plants serving subdivisions.

To determine if the plant is receiving excessive flows (hydraulic surges), the operator should observe the flow over the weir in the clarifier during peak-flow periods. Flow should be uniform over the entire length of the weir, and the quality of the effluent should remain the same. If the plant is operating properly, the effluent should be clear. If solids begin to appear at the overflow weir and the rise of liquid level in the clarifier exceeds 12 mm (0.5 in.), then hydraulic washout could be occurring. Clarifiers are designed to allow the solids to settle and the clear liquid to flow gently over the weir. If hydraulic surges are washing solids from the plant, the solids are lost during peak flows and the plant should be evaluated by the owner, local regulatory agency, or design engineer.

Problems associated with hydraulic surges (peak flows) can be eliminated by installing an equalization tank before the package wastewater treatment plant. Hydraulic surges flow into the equalization tank and are stored. The wastewater can then be pumped to the aeration chamber at a uniform rate of flow. Because the flow to the equalization tank (often referred to as a *surge tank*) during periods of peak flow will exceed the pumping rate to the aeration chamber, the liquid level in the equalization tank will rise. Sufficient storage must be provided in the surge tank to allow for this rise. Various factors will determine the required storage capacity. These design criteria will vary depending on the region in which the plant is located. However, not all plants need equalization tanks. For those plants that do, the volume of the equalization tank may be from 5 to 60% of the aeration chamber volume.

Wastewater is typically removed from the equalization tank with constant speed centrifugal grinder pumps. Pumps installed in the equalization tank must be able to pass the solids found in raw wastewater. Although some manufacturers use air ejectors to remove the wastewater from the equalization tank, submersible grinder pumps are also used. If other types of pumps are used, a bar screen must be installed at the influent to the equalization tank. Because the water level in the surge tank changes, the pumping rate of a constant speed centrifugal pump will vary. As the water level is pumped down in the equalization tank and the pumping head increases, the pump will deliver less wastewater to the aeration tank.

A cost-effective way to overcome the problem of the changing level is to install a constant head pump chamber within the flow equalization tank, as illustrated in Figure 3.1. Submersible pumps in the flow equalization tank

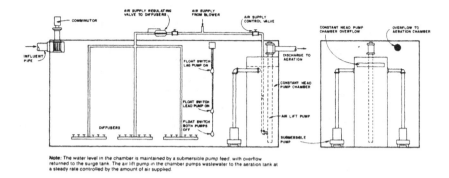

Note: The water level in the chamber is maintained by a submersible pump feed, with overflow returned to the surge tank. The air lift pump in the chamber pumps wastewater to the aeration tank at a steady rate controlled by the amount of air supplied.

Figure 3.1   Constant pump head pump chamber within flow equalization tank.

feed the airlift pump chamber, keeping it filled to a constant overflow level near the top. The overflow is returned to the flow equalization tank.

The vertical distance from which the wastewater must be lifted from the chamber's overflow level to the aeration chamber remains constant, no matter what the liquid level in the flow equalization tank. With a constant low lift now required, an air lift pump works well in the constant head chamber, just as it does in the clarifier to return settled sludge to the aeration chamber. The clarifier is, in fact, a constant head pumping chamber, because the liquid level in the clarifier is maintained by the overflow weir.

An overflow from the equalization tank to the aeration chamber should be provided as an emergency bypass, should both submersible pumps fall or the constant head air lift pump malfunction. Many regulatory agencies require a gravity overflow.

*W*ASTEWATER CHARACTERISTICS

HIGH-STRENGTH WASTEWATER. Package plants provide wastewater treatment services for small communities, parks, recreational facilities, industrial facilities, boating marinas, commercial areas, recreational vehicle parks, and other single-user-type facilities. Wastewater from these sources can be of a higher strength than typically found.

BIOCHEMICAL OXYGEN DEMAND. The biochemical oxygen demand (BOD) test is the most widely used parameter of organic pollution applied to both wastewater and surface water. The test simply measures the amount of dissolved oxygen used by microorganisms (seed) in the biochemical oxidation of organic matter over a five-day period. The BOD test is significant because it is used to determine the approximate quantity of oxygen that will be required

---

to stabilize the waste at a wastewater treatment plant. Typical domestic wastewater has a BOD of approximately 200 mg/L, whereas high-strength waste can have BODs in the range of 1000 to more than 10 000 mg/L.

CHEMICAL OXYGEN DEMAND. The chemical oxygen demand (COD), like the BOD, is used to measure the content of organic matter in wastewater. Unlike the BOD, the COD uses chemicals to determine the oxygen requirement of waste. This allows the COD to be used to determine the organic matter in waste that is toxic to biological life. The COD of waste is generally higher than the BOD because more compounds can be chemically oxidized than can be biologically oxidized. For many wastes, there can be found a consistent correlation between the BOD and COD, which is very useful because the COD test is run in a few hours, whereas the BOD test is run over a five-day period. Domestic wastewater typically has a COD of 500 mg/L and, like the BOD, high-strength wastewater can have COD in the range of 1000 to more than 10 000 mg/L.

TOTAL SUSPENDED SOLIDS. The most important physical characteristic of wastewater is its total solids content. This is the amount of solid matter that is one micron or larger in size and generally is the matter that can be removed by settling. The remainder of the solid matter is either dissolved or colloidal. Typical wastewater has a total suspended solids (TSS) level of approximately 200 mg/L, whereas high-strength waste can have TSS in the range of 1000 to 10 000 mg/L.

A comparison of wastewater strengths from nonmunicipal sources is shown on Table 3.1.

*A*DDITIVES

Additives are chemicals added to boat and recreational vehicle holding tanks and portable toilets to reduce or eliminate offensive odors created by waste disposed in these units. Over the years, the types of additives available have changed. Zinc salts were once very prevalent, but now are not available because of their toxic nature to the environment. California prohibited the sale or use of zinc additives in 1978. Today, the prevalent additives are those containing chemical combinations of formaldehyde, enzymes, dyes, and perfumes. Formaldehyde-type additives are used to stop any biological activity in the wastewater and thus prevent odors. The formaldehyde is supposed to kill any biological microorganisms in the waste that generate odor-causing compounds as they typically work. Enzyme additives work the opposite way, by promoting biological activity to liquefy the wastes and eliminate odors. In this case, the enzymes (provided to treat the waste) do not generate odor-causing compounds.

Table 3.1   Comparison of wastewater characteristics from various sources (Jenkins et al., 1993).

Source location	BOD (mg/L)	COD (mg/L)	Soluble COD (mg/L)	TSS (mg/L)	VSS[a] (mg/L)
Oregon State[b] Marine Board study average	2990	8020	5140	1600	1370
Federal Register (1994)[c]	1700 to 3500				
Robins and Green (1974)[c]	2710	6180		2860	2310
Brown et al. (1984)[d]	3110	8230	2930	3120	2460
Pearson (1980)[d]	3080	6210		3850	3330
Sealand (1990)[d]	3000 to 6000	15 000 to 18 000		13 000 to 15 000	
Brestad et al. (1971)[d]	1840 to 7590	5600 to 22 000		1120 to 20 500	1020 to 18 400

[a]VSS = volatile suspended solids.
[b]Results from 1994 Oregon State Marine Board study.
[c]Sample from study of recreational boat waste.
[d]Sample from study of recreational vehicle waste.

TOXICITY EFFECTS ON TREATMENT PLANTS

A study by the Maryland Department of the Environment (Buchart-Horn, Inc. and Versar, Inc., 1992) recommended dilutions for high-strength waste containing various additives, based on various treatment processes. The additives included formaldehyde and paraformaldehyde; detergents, cleaning agents, and disinfectants (quaternary ammonium compounds [QACs]); and zinc compounds. The treatment processes included the activated sludge and trickling filter processes, anaerobic digestion, and septic tanks. Additives containing formaldehyde and paraformaldehyde required the greatest dilution for the activated sludge, trickling filter, and anaerobic digestion processes. Dilutions for zinc compounds are not discussed here because they are no longer used.

For municipal wastewater treatment facilities that use the activated sludge process and treat high-strength waste containing formaldehyde and paraformaldehyde, the Buchart-Horn, Inc. and Versar, Inc. study recommends that the waste be diluted 7 to 1 with domestic waste. This would lower the additive concentration from 400 mg/L (concentration found in high-strength waste) to 50 mg/L (concentration that would not affect treatment process).

Table 3.2   Volume of boat waste affecting plant toxicity for various treatment systems.

Plant flowrate, L/s (gpd)	Activated sludge or trickling future, L/s (gpd)	Anaerobic sludge digestion, L/s (gpd)
0.44 (10 000)	0.06 (1400)	0.03 (770)
2.2 (50 000)	0.3 (7000)	0.2 (3850)
4.4 (100 000)	0.6 (14 000)	0.34 (7700)
22 (500 000)	3 (70 000)	1.7 (38 500)
44 (1 000 000)	6 (140 000)	3.8 (77 000)

Table 3.2 shows the maximum volume of boat waste a wastewater treatment plant could accept and not exceed the dilution recommendations for boat waste additives. The table is based on (1) the wastewater treatment process used, (2) the treatment plant flowrate, and (3) the previously discussed dilution requirements.

For example, an activated sludge plant treating approximately 2 L/s (50 000 gpd) of wastewater could only accept approximately 0.3 L/s (7000 gpd) of high-strength waste and not exceed the 7 to 1 dilution recommendation. For anaerobic digestion, the flowrate is that into the digester and not the plant wastewater flowrate. Sludge flowrates are typically less than 1% of the plant wastewater flowrate.

The reader is referred to the Buchart-Horn, Inc. and Versar, Inc. study (1992) for more detailed information on toxicity.

OPERATION OF EQUALIZATION TANK

In a typical plant equalization tank, two submersible pumps, working alternately, are used to keep the constant head pump chamber filled to overflowing. The pumps are controlled by three mercury-type float switches and an alternator to give both pumps equal run time. The top float switch will turn on the lag pump and activate a high water alarm if the lead pump fails to turn on. The next float switch, set well below the top one, will turn on the lead pump and air lift blower when the equalization tank becomes approximately 25% full. The level at which the lead pump is set to turn on can vary, but generally should be low enough so that the water level rarely reaches the top float switch.

The blower supplying air to the air lift pump is interlocked electrically (each turns on and off at the same time) to the lead submersible pump in the equalization tank. The lead pump and blower will continue to operate until the equalization tank is pumped down to the low-level float switch, which turns off both the pump and blower. This float should be located several centimeters (inches) above the submersible pumps' intakes.

A valve on the air supply line to the air lift pump allows the operator to control the pumping rate to the aeration chamber by the amount of air that is injected. A second air supply line is located in the equalization tank, which directs air to a set of diffusers that provide mixing and preaeration of raw wastewater. A pressure regulator is required on this air line to prevent all the air from going to the diffusers as the submersible pumps lower the liquid level in the surge tank. A regulator, set at 34 kPa (5 psi), provides sufficient resistance to prevent excessive loss of air through the diffusers.

Air may be provided to the diffusers in the equalization tank from one of two sources, depending on the mode of operation. The blower supplying air to the constant head lift pump may provide air if there are no prolonged periods without flow. In this case, the equalization tank is aerated only when the submersible pumps run, because that is the only time the air lift pump operates. This operational mode is satisfactory when flow is received daily and the submersible pumps operate frequently. If, however, flows are received for only five days a week, such as at a school, or only on weekends, such as at a campground, the raw wastewater will be better aerated by supplying air to the diffusers from blowers that provide air for the wastewater treatment plant. These blowers are typically controlled by time clocks and operate on a regular daily basis. This operational mode prevents raw wastewater from becoming septic over long periods of no flow.

*A*IR LIFT PUMP

Flow from the equalization tank is typically controlled by an air lift pump or flow control box. An air lift pump is a simple device to lift water using compressed air. As illustrated in Figure 3.2, the pump consists of a 50- to 100-mm (2- to 4-in.) diameter vertical pipe (eductor) with a side discharge (tee) above the water level at the top of the pipe. The vertical distance from the water level to the point of discharge is called the *lift*. Compressed air is injected approximately 150 mm (6 in.) from the bottom of the eductor. The bottom of the eductor (tail pipe) should be located 150 mm (6 in.) off the bottom of a constant head pumping chamber, which may have a hoppered bottom. This keeps wastewater in the bottom of the chamber moving toward the eductor fast enough (scouring velocity) to keep solids from accumulating on the bottom of the chamber.

The distance from the water level down to the point at which air is injected is called the *submergence*. The air lift pump operates most efficiently when the ratio of submergence to lift is high, for example, 9 to 1. The efficiency of an air lift pump drops off as this ratio decreases, because more air must be supplied to maintain the same pumping rate. If the ratio gets too low, so much air would be required that the air lift pump becomes impractical.

What makes the air lift pump work? The injected air bubbles make the water inside the eductor lighter (less dense) than the water outside. The relatively heavier (more dense) water outside the eductor generates enough

---

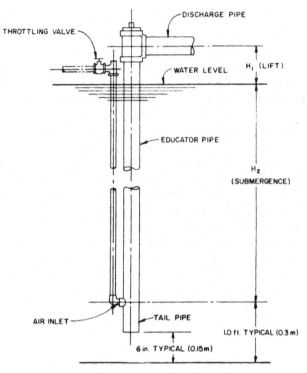

THROTTLING VALVE

DISCHARGE PIPE

WATER LEVEL

H₁ (LIFT)

EDUCATOR PIPE

H₂
(SUBMERGENCE)

AIR INLET

TAIL PIPE

1.0 ft. TYPICAL (0.3 m)

6 in. TYPICAL (0.15 m)

Note: An air lift pump gets lifted by injecting air near the bottom of the eductor pipe. If the submergency lift ratio is high enough (8.5 to 1 shown here), the more dense water outside the eductor will push the air-water mixture up the eductor to the discharge point. At constant submergence lift air supply, the flow will remain constant.

Figure 3.2   Air lift pump.

head (pressure) to push the air-liquid mixture up the eductor. This is why so much submergence is required to get such a small lift. A valve on the air supply line is used to control the pumping rate.

FLOW PROPORTIONING CHAMBER

A less expensive, but also less satisfactory, way to overcome the problem of a changing level in the flow equalization tank is with a flow proportioning chamber because no air supply is required. Mounted above the flow equalization tank and water level of the aeration chamber, the box receives wastewater from a submersible centrifugal pump in the flow equalization tank. The box has an adjustable V-notch weir that allows control of flow by gravity to the

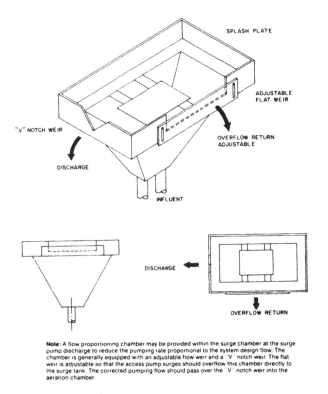

SPLASH PLATE

ADJUSTABLE
FLAT WEIR

"V" NOTCH WEIR

OVERFLOW RETURN
ADJUSTABLE

DISCHARGE

INFLUENT

DISCHARGE

OVERFLOW RETURN

Note: A flow proportioning chamber may be provided within the surge chamber at the surge pump discharge to reduce the pumping rate proportional to the system design flow. The chamber is generally equipped with an adjustable flow weir and a "V" notch weir. The flat weir is adjustable so that the access pump surges should overflow this chamber directly to the surge tank. The corrected pumping flow should pass over the "V" notch weir into the aeration chamber.

Figure 3.3   Flow proportioning chamber.

aeration chamber. The overflow is returned to the flow equalization tank. Figure 3.3 illustrates a flow proportioning chamber.

One drawback to the flow proportioning chamber is that the submersible pump will deliver more water to the box when the liquid level is high in the surge tank, and will deliver less as the tank is pumped down. Consequently, no matter where the V-notch weir is set, the setting will be too low initially and too high at the end to maintain the same flow to the aeration chamber throughout the pump down the equalization tank. Despite this drawback, a flow control box can equalize the hydraulic surge to some extent.

EQUALIZATION TANK OVERFLOW—PERMIT REQUIREMENTS

The U.S. Environmental Protection Agency requires 100% standby pumping capacity in the equalization tank, which is provided for by the two submersible pumps. However, flow interruption would occur during periods of power failure; therefore, installation of emergency overflow by gravity from the equalization tank to the aeration tank is recommended.

GENERAL MAINTENANCE

A fixed skimming device that removes grease balls and scum in the equalization tank is impractical because the water level changes continuously. The operator must remove these floating materials manually with a wire mesh, such as ordinary window screen. A simple device that is convenient for removing floating material can be made readily by mounting the mesh on a wire frame that has been fitted with a long handle. The same kind of maintenance is generally required in the clarifier. Material removed from the flow equalization tank should be put into a lined trash can with a tight-fitting lid. The best method of final disposal is at an approved landfill.

SUMMARY

(1) Flowrates to treatment plants vary substantially during the day and week.
(2) Flow equalization is a method of temporarily storing wastewater during periods of excessive flows, then feeding it to the plant uniformly.
(3) High-strength wastewater, with a BOD in the range of 43 830 to more than 438 300 L/s (1000 to more than 10 000 mgd), can cause problems to package plants.
(4) Additives in wastewater from recreational vehicles and boat holding tanks can be toxic to package plants.

REFERENCES

Brown, C. A.; Kiernan, K. E.; Ferguson, J. F.; Benjamin, M. M. (1984) Treatability of Recreational Vehicle Wastewater in Septic Systems at Highway Rest Areas; Transportation Research Record 995; Transportation Research Board: Washington, D.C.

Buchart-Horn, Inc.; Versar, Inc. (1992) A Survey of the Quantity, Characteristics, and Potential Impacts of Boat Pumpout Waste Generated Within the Chesapeake Bay Region of Maryland; Report to the State of Maryland Department of the Environment; Marina Sewage Treatment Survey Project MDE-90-2.2 WQA; Annapolis, Maryland.

Oregon State Marine Board (1995) Effects of Boat Waste Disposal at Municipal Wastewater Treatment Facilities.

Pearson, F.; et al. (1989) Report to the Virginia Department of Health on Effects of Holding Tank Additives on Treatment of Boat Holding Tank Wastes; Virginia Water Resources Research Center.

Robins, J. H.; Green, A. C. (1974) Development of On-Shore Treatment System for Sewage from Watercraft Waste Retention System; U.S. Environmental Protection Agency, EPA-670/2-74-056.

Sealand Technology, Inc. (1990) Internal Memo from Fred Morris, Health Hazards Related to Servicing of Marine & RV Sanitation Systems.

U.S. Department of the Interior (1994) Clean Vessel Act: Pumpout Station and Dump Station Technical Guidelines. *Fed. Regist.*, March 10; Fish and Wildlife Service.

Chapter 4
Protective Devices

INTRODUCTION

Protective pretreatment devices are designed to remove nontreatable material from wastewater before it reaches the aeration chamber or precondition wastewater and make it easier to treat in the aeration chamber. Most large treatment plants are designed to allow wastewater to flow through a primary clarifier for removal of settleable or floatable materials. This primary settling or treatment typically removes grit, sand, and grease, and reduces organic loading on the aeration chamber by approximately 30%. Small package wastewater treatment plants do not have separate units for primary treatment; instead, they combine primary and biological treatment. Small plants have protective devices similar to those of larger plants. Many regulatory agencies consider the equalization tank a pretreatment device. Equalization was discussed as a separate topic in Chapter 3 because of the numerous concepts its operation involves.

BAR SCREENS

Bar screens are simple, nonmechanical devices found on most package plants. Fabricated of metal bars or rods, they are designed to catch large objects in raw wastewater.

Bar screens should be fabricated from 12- × 50-mm (0.5 × 2 in.) bars rather than 12-mm-diameter (0.5-in.-diameter) rods. This eliminates the need for transverse stiffeners, permitting easy cleaning of the screen. If the bar screen is located in the aeration chamber, the bottom one-third should be under water. The rolling action in the aeration chamber will break up large organic solids, thereby preventing buildup of treatable solids on the bar screen. The top of the bar screen should be well below the inlet pipe so that solids do not block the inlet pipe and cause wastewater to back up in the line. Some manufacturers locate the skimmer and sludge return discharge into the bar screen, thus capturing materials that initially pass through it.

Although most bar screens installed on small package wastewater treatment plants are similar to those in Figure 4.1, their location and design have caused some problems. The bar screen is designed to remove large untreatable material from the wastewater flow and should be located as shown in Figure 4.2. Most manufacturers place the comminutor before the bar screen; this causes nontreatable material to be ground into small pieces and passed through the bar screen.

The bar screen should be checked during each visit and cleaned as necessary. This can be done with a rake designed so that its tines will mesh with the bars of the screen. A drying rack is recommended to drain the screenings before disposal. Material should be removed from the drying rack after draining and placed in a lined trash can with a tight-fitting lid to avoid odor and insect problems. If a drying rack is not provided, an industrious operator could drill holes in the bottom of an approximately 19-L (5-gal) pail and suspend a bag of netting inside to dewater the screenings. The best method of final disposal is at an approved landfill or, if approved by the local regulatory agency, on-site burial.

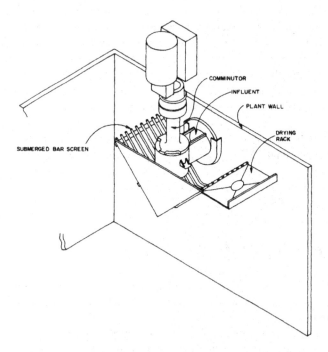

Figure 4.1   Bar screen and comminutor (common design).

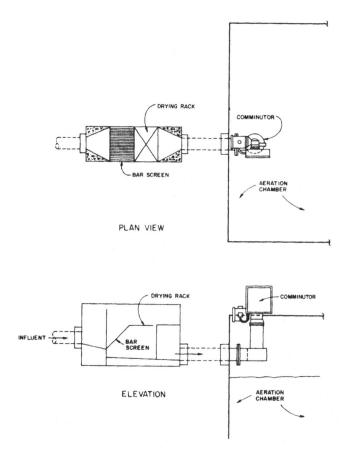

Figure 4.2   Bar screen and comminutor (improved design).

COMMINUTOR

As illustrated in Figure 4.3, the comminutor or grinder is a revolving drum with teeth that pass through stationary cutter combs. Comminutors should have an automatic reverse to dislodge large objects that cannot be shredded. The comminutor is designed to shred solids into smaller, more readily digestible particles. The comminutor runs continuously and, on most small plants, is located in front of or parallel to the bar screen.

The two major disadvantages of using a comminutor are power consumption and entry of nontreatable waste into the plant. Because the comminutor is often located ahead of the bar screen, nonbiodegradable waste such as fibrous material, sticks, and plastics are shredded and passed through the bar screen. Ideally, the bar screen should precede the comminutor, even if this increases the initial cost of the plant. The maintenance time saved by keeping nonbio-degrable material out of the plant is well worth the added expense. With the

---

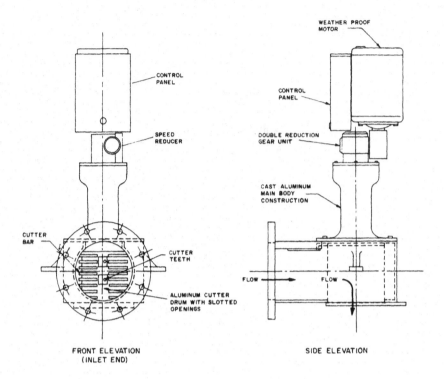

FRONT ELEVATION
(INLET END)

SIDE ELEVATION

Figure 4.3 Typical comminutor.

comminutor ahead of the bar screen, all the advantages of the bar screen are
lost. Most plants will actually require less maintenance with just a bar screen
than with a bar screen preceded by a comminutor.

Normal maintenance of the comminutor includes sharpening the cutter teeth
and inspecting to ensure that nothing is caught in the unit. **Caution: Never
use hands to remove objects stuck in the cutter teeth,** even when there is
no power to the unit. Work on a comminutor should never be done unless the
power is off and appropriate lockout-tagout procedures are followed. Cutter
teeth and combs must be replaced when worn. A drive box may require grease
or regular oil changes. Finally, the drive motor should be serviced according
to the manufacturer's service manual.

*T*RASH *TRAP*

This pretreatment unit is not common to most package treatment plants, but is
required by some states. It operates similarly to a septic tank; floatable materials,
such as grease and oil, are allowed to rise to the surface, while heavy solids,
such as sand and grit, settle to the bottom. The trash trap is a tank, baffled at
the inlet and outlet, with a volume that is approximately 10% of the aeration

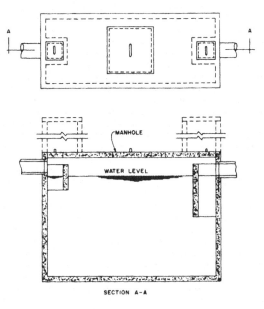

Figure 4.4   Typical trash trap.

chamber. It is nonmechanical and therefore requires little regular maintenance. The outlet baffle should extend to 40% of the liquid depth to retain floating solids. Figure 4.4 illustrates a typical trash trap.

The biggest disadvantage of a trash trap is that material retained in the unit will become septic, which could result in a septic package plant influent. The trash trap may also result in an underloaded wastewater treatment plant. Foul odors are also frequently associated with this unit. Consequently, trash traps are not recommended as a pretreatment device. If grease or oil is a problem, a grease trap can be installed locally on the sewer generating the problem.

Trash traps must be routinely checked for solids accumulation. Pumping is required when approximately 40% of the volume is occupied by floating and settleable solids. The contents should be removed by a licensed septic tank pumper and disposed of at a nearby municipal facility or other approved location. **This material must not be pumped into the package plant.**

GREASE TRAP

Where high grease loads are anticipated in raw wastewater, such as in restaurants, school kitchens, etc., a grease trap may be installed. A grease trap operates on the same basic principle as the trash trap. The primary difference is that the grease trap is smaller, with the effluent baffle extending almost to the bottom of the tank so that only floatable materials are captured. Figure 4.5 shows a typical grease trap. Excessive grease can cause maintenance problems

---

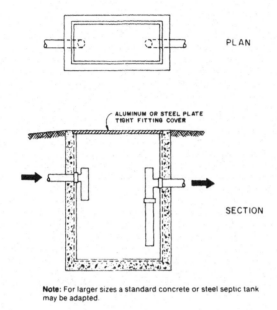

PLAN

ALUMINUM OR STEEL PLATE
TIGHT FITTING COVER

SECTION

Note: For larger sizes a standard concrete or steel septic tank
may be adapted.

Figure 4.5   Typical grease trap (600 L).

in the plant, including foaming, poor settling characteristics, and excessive
scum. Some states require grease traps in such instances. Improvements in
plant operation will more than justify the additional cost of the grease trap.
They typically provide a detention time of 30 to 60 min for the wastewater.
Garbage disposals (grinders) should be discouraged, especially on systems
with grease traps.

Maintenance of a grease trap is similar to that of a trash trap. The effluent
baffle should be checked routinely to be sure that it is intact. The tank should
be located for quick and easy cleaning. The location should be as close to the
source as possible to prevent the grease from congealing inside the pipe, yet
not so close as to cause hot effluent to carry grease beyond the trap. The
location of the trap should be outside the structure for easy maintenance. An
inconvenient location will contribute to neglect; frequent cleaning—perhaps
every 2 to 4 weeks—may be needed for optimum performance. Disposal of
wastes from a grease trap may be a problem because some rendering companies and municipal plants do not accept this type of waste. The appropriate
regulatory agency can be consulted regarding disposal of grease trap wastes.

*E*LECTRICAL PROTECTION

A *phase monitor* is a device to protect a three-phase electric motor from failure
because of high peaks in voltage or a loss of a phase. During an electrical

storm or power failure, one or more legs of the three-phase power may be lost. Without a phase monitor to shut down the system, an electric motor would continue to operate, resulting in overheating and short-circuiting.

Phase monitors should be used for lift stations and wastewater treatment plant motors operating on three-phase power and subject to lightning or power outages. Operators can find the phase of each electrical motor on the nameplate.

Chapter 5
The Aeration Chamber

INTRODUCTION

After raw wastewater flows through the various protective devices where pretreatment occurs, it flows to the aeration chamber or basin. The aeration chamber is the key part of the secondary treatment process; bacteria and other microorganisms thrive and multiply there as they consume food (organic material) in the wastewater. This aerobic biological process (biological in that bacteria and other microorganisms are essential to the process and aerobic in that these microorganisms need air or dissolved oxygen to survive) is called an *activated sludge treatment system*. The microorganisms, or microbes, are typical of all living things in that they need food, oxygen, and a compatible environment in which to thrive and multiply.

The aeration chamber houses the microorganisms, and most of the biological action occurs in the aeration chamber. The clarifier provides the opportunity for the microorganisms to separate from the treated wastewater and settle, thus producing a clear effluent discharge. Figure 1.1 shows the basic parts of a package extended aeration wastewater treatment plant (WWTP).

The aeration chamber is the key to secondary biological treatment. *Secondary treatment*, as the term is used here, is the biological conversion of nonsetteable solids into a form that will settle. Bacterial use organic materials in wastewater as a food source for growth and reproduction. In doing so, the organic pollutants become bacteria. Higher life-form microorganisms eat the bacteria and the food chain moves upward; this is an ongoing process. The mass of organisms in the biological system (called *biomass* or *mixed liquor*) contains all kinds of life forms, from bacteria to worms. The term for the process that uses this mass of suspended microbes is *activated sludge*, and the specific mode of activated sludge that this text deals with is the *extended aeration mode*. While there are unique characteristics in the extended aeration mode, the various modes of the activated sludge process also have many common characteristics.

THE ACTIVATED SLUDGE SYSTEM

Activated sludge systems have four essential units that make up the total treatment process. These units are (1) the aeration tanks with their aerating and mixing abilities, (2) the secondary clarifiers with solids-settling environ-

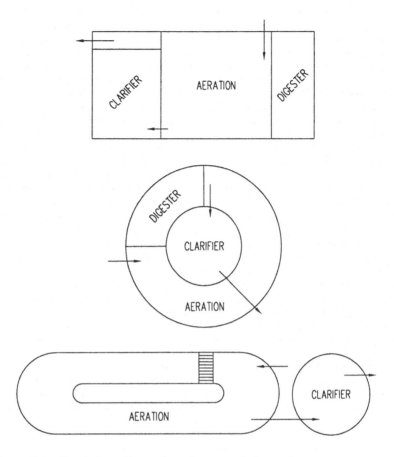

Figure 5.1   Typical configurations for extended aeration.

ments and sludge withdrawal conditions or mechanisms to remove the settled activated sludge, (3) pumps or other methods for returning settled activated sludge to the aeration tanks, and (4) means of removing a portion of the solids from the system.

The aeration tanks are vessels, which are sized for a given activated sludge system and flow. They provide a home for the bacteria to eat, grow, and reproduce. They provide the appropriate amount of time for the bacteria to complete these functions. These tanks must be aerobic (that is, have a surplus of dissolved oxygen) at all times.

Oxygen is typically provided in aeration tanks by one of two systems: (1) a diffused air system or (2) a mechanical air system. In a diffused air system, compressed air is blown into the aeration tank contents and diffused as tiny bubbles. As the bubbles rise, oxygen from the bubble transfers to the liquid and becomes available for use by the microbes. A mechanical air system consists of a system that breaks the water into droplets and thereby increases their exposure to the atmosphere. Oxygen is dissolved into the droplet from the point where the water meets the air. To some degree, oxygen is pushed into the water for transfer near the surface. In a mechanical aeration system, dissolved oxygen levels can be controlled by (1) aerator submergence or (2) aerator speed.

The secondary clarification tanks provide a quiescent area for biological activated sludge solids (or floc) to separate from the mixed liquor coming from the aeration tanks. These tanks may also be equipped with skimming devices to remove any floating materials from the surface of the water in the tank.

As activated sludge microorganisms continue to grow and multiply, some of them must be removed to keep the system in balance. If they are not removed, the system will lack enough food or oxygen to support them. This will eventually cause many unwanted plant problems such as poor settling, solids in the plant effluent, and excessive flows to solids handling units. These excess activated sludge solids, called *waste activated sludge* (WAS), must be removed or "wasted". It is important to the plant operation that the wasting rate be sufficient to maintain just enough activated sludge solids in the aeration system to properly treat the incoming wastewater. The operator should note that the means of wasting solids is not the major concern; the primary consideration is the maintenance of the most effective biological solids balance in the aeration system.

Because unique and varying systems have varying resources, this chapter is structured to address two extremes of operation. First, there will be a detailed description of operation. This is done for systems with heavier staffing capabilities and more comfortable operational budgets. More importantly, it is done because all operators should understand the basics of the activated sludge process when operating an activated sludge plant. Even though there are tips and procedures that get the operator started, the goal should be to become as familiar with the process as possible. The second approach in this chapter is a more basic or rudimentary approach. The reality of package plant operation is that is typically performed on an extremely slim budget, and the responsibilities are given to a relatively untrained operator. That is where this text comes in. The intent of the formatting is to provide an exposure to the principles of activated sludge while providing practical operational tips. As familiarity with the process grows, the more specific activated sludge text may be referred to again.

The chapter first addresses the physical components of an extended aeration package plant, then the description of the activated sludge process. Finally, it discusses process control.

AERATION TANK

The aeration tank performs multiple functions; it provides dissolved oxygen for microorganisms, mixes raw wastewater with the mixed liquor or biomass, and provides time for the biology of wastewater treatment to take place. The oxygen supply must be sufficient to maintain a minimum dissolved oxygen concentration of 2.0 mg/L in the aeration chamber at all times. If an inadequate amount of air is supplied to the aeration chamber, the aerobic bacteria, those that do the vast majority of the treatment work, will die. Anaerobic bacteria (those that live without free oxygen) will begin to grow, producing disagreeable hydrogen sulfide or rotten egg odors. More serious process problems will soon follow. A third type of bacteria, called facultative, can live under either condition.

The aeration basin must be kept uniformly aerated and mixed, thereby providing oxygen for the bacteria, keeping the solids in suspension, and allowing a rapid mix of raw wastewater with bacteria for oxidation and synthesis of the organic matter. The aeration basin should be inspected each time a site visit is made to determine that uniform mixing is occurring. Improper mixing in the aeration chamber could result in sludge deposits that may become septic, thus hindering proper operation of the aeration chamber. As a rule for most domestic facilities, if mixing is by diffused air, and mixing is adequate, aeration will also be adequate.

SIZING. Standard hydraulic retention time (the time the wastewater spends under aeration) for an extended aeration process is 24 hours. This is a value calculated without taking any recycle flows into consideration. The 24-hour aeration period is the biggest process difference there is between the conventional activated sludge process and the extended aeration modification of the process.

CONFIGURATION. On most small extended aeration package treatment plants, the aeration chamber is rectangular, with the longer side being four to five times longer than the width. These are typically aerated by diffused air. Some units are square, with a mechanical aerator located in the center. This is termed *complete-mix* because the wastewater enters and, in a short time, because of its central positioning, the aerator completely mixes the wastewater with the mixed liquor. Another common configuration is the oxidation ditch. It is typically a relatively shallow tank, 2 to 2.5 m (6 to 8 ft) deep, shaped like a racetrack. Aeration and mixing are by mechanical brush rotors. From the aeration tank, wastewater flows into the clarifier through a baffled crossover located at the opposite end. Figures 5.1a, b, and c show typical configurations for various package extended aeration plants.

AERATING AND MIXING CAPABILITY

DIFFUSED AIR. Oxygen, in the form of diffused air, is supplied to the aeration chamber from the blowers to vertical drop pipes that are typically located along the longer sides of the chamber. At the bottom of the pipe, it typically tees off into horizontal pipes or laterals that sit a few feet off the bottom. The pipes are connected to, and in many cases are, the diffusers. In the cases where they are the diffusers, the lateral pipe is simply supplied with holes drilled into the bottom of the pipe. In more elaborate systems, pieces of equipment manufactured specifically to make the air diffuse and create smaller bubbles are installed on the laterals. Each drop pipe has a plug valve so that the rate of airflow can be adjusted. Figure 5.2 shows a typical diffuser assembly. This arrangement causes a wall of air bubbles to rise on one side of the tank. As the bubbles rise, they lift wastewater from the bottom of the tank toward the top. The wastewater lifted is replaced by mixed liquor from below. This air-lifting results in the mixed liquor flowing across the water surface. This produces a rolling action which, when combined with incoming flow, causes the wastewater in the aeration tank to travel a cork-screw path from end to end. This action provides excellent mixing and gives the bacteria an opportunity to obtain both food and oxygen necessary to thrive.

The air piping should be below the plant grating or otherwise safely located to prevent a tripping hazard. A check valve should be installed on the discharge side of all blowers to reduce diffuser plugging from solids that flow into the diffusers when the blowers are off. This will increase the time between cleaning the diffusers and downfeed pipes. The check valve is also necessary to prevent

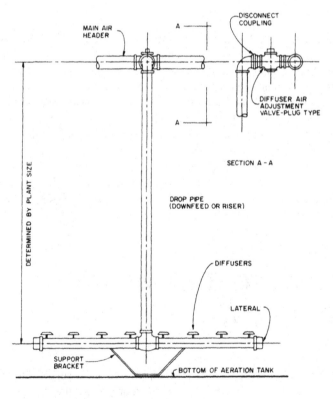

Figure 5.2 Typical diffuser assembly.

the standby blower from being forced to run backwards by the operating blower's discharge.

Each air downfeed pipe should have a plug valve with a mechanical union disconnect on the pipe on the diffuser side of the valve to allow removal and servicing of downfeed pipes while the blowers are running. Plug valves are recommended because they require little maintenance, do not vibrate open or shut, allow visual evaluation of their position without having to remove the grating, and are accurate enough for use in an aeration tank as throttling valves. When maintenance or cleaning is required, the downfeed pipes should be tied off and then removed carefully, because their support(s) will not usually prevent them from falling into the aeration basin when disconnected. Plug valves will generally be wide open unless an adjustment is necessary to balance the roll above each downfeed pipe. Balancing is done when all downfeed pipes and diffusers are clean and intact. Diffusers must be cleaned periodically. Cleaning will require the removal of the diffusers, and care should be taken when reinstalling them. If a diffuser is off, most of the air will escape through that opening, creating an uneven rolling action. This condition will resemble a boiling action at the water surface. The operator should check each downfeed pipe to ensure it is plumb. If the downfeed pipe is not plumb, the lateral will not be level; this will cause some of the diffusers to be higher than others. More air will escape from the high diffusers, creating an uneven rolling action. The length of all downfeed pipes should be the same; if they are not, more air will escape from the higher header, resulting in an uneven rolling action. To ensure uniform mixing in the aeration chamber, diffusers must be clean, laterals must be level, downfeed pipes must be the same length, and all diffusers must be on. Air pipes and valves may be checked for leaks with soapy water. If there are no air leaks and there appears to be an insufficiency of aeration, the motor, blower, sheaves sizes, and revolutions per minute (rpm) should be checked to determine whether the required air volume is actually being supplied.

There are certain things at a WWTP that are there to stay, and the operator has little or no ability to change them. The tanks themselves are good examples. Nevertheless, there are major items supplied at construction that should be checked in attempts to prevent problems from occurring down the road. The operator should consult the owner's manual to ensure that the proper-sized motors and blowers have been installed on the plant. If an owner's manual is unavailable, the operator should measure the length, width, and depth of the aeration chamber. These measurements will determine the volume of the tank. As stated earlier, the extended aeration process uses a 24-hour detention time. This means that the volume is also the design capacity of the plant. The design flow capacity allows the manufacturer to be able to tell the proper size blower and motor for the plant.

The amount and rate of air pumped by the blower is determined by the rpm of the blower. Typically, the blower and motor are adequately sized to provide the aeration required at design. The manufacturer should be able to provide a curve that illustrates the airflow at various speeds. The operator should keep in mind that speeding up a blower shortens its life in almost every situation. The operator needs to make certain that insufficient air is, in fact, the real problem before increasing speed. If the operator does have good reason to believe that additional aeration is needed, the following conditions should be considered.

The electric motor used to power the blower is designed to operate at a specified constant rpm. Consequently, the speed at which the blower turns is determined by the size of the sheaves or pulley on the blower as compared to the one on the constant speed motor. Figure 5.3 shows a typical blower–motor arrangement.

By comparing the diameter of the two sheaves, the rpm of the blower may be determined. By varying the size of the sheaves, manufacturers can use the same size blower and motor on several sizes of WWTPs. The number of rpm produced by the electric motor is given on the motor name plate. The rpm of the blower can be calculated by dividing the diameter of the motor sheaves by the diameter of the blower sheaves and multiplying this number by the motor's rpm. The diameter of the sheaves is the measurement from one side of the sheaves to the opposite side, making sure to measure through the center shaft. On most small extended aeration plants, the sheaves are not adjustable. Changes, if necessary, must be made by changing the size of the sheave on the blower or motor (or both). A portable tachometer may be used to check the rpm of the motor or blower.

The importance of the proper amount of air in the aeration chamber cannot be overemphasized. Many factors that affect the dissolved oxygen concentration and mixing in the aeration chamber have been noted. In later chapters, additional factors that affect the air supply to the aeration chamber will be

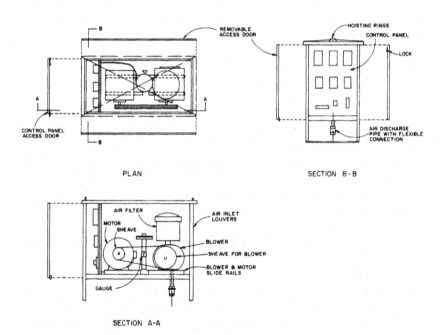

Figure 5.3 Typical blower–motor assembly.

discussed. With experience, the operator will be able to determine which factors are causing operation problems. In seeking solutions to a problem, always try the obvious or simplest solutions first; one should not rebuild a plant if only a valve adjustment is required.

BLOWERS AND MOTORS. Two blowers and motors, each with sufficient aerating capacity to supply all air needs, are typically installed in a package plant. This arrangement extends blower life by allowing for alternate use, and provides 100% standby in the event of an emergency.

There are several types of low-pressure blowers for small package extended aeration plants. The manufacturer's operating manual should indicate the type and size of the blower installed in the plant and give all maintenance procedures. If a manufacturer's manual is unavailable, the blower nameplate will provide all information required to identify the blower. The manufacturer should then be contacted to obtain all maintenance information. Blower maintenance such as changing oil and grease is important to successful operation. The operator should determine what maintenance is required for the blower and follow the manufacturer's recommendations exactly. The blower is a critical element in the operation of the treatment plant.

Blowers are sized according to two criteria: to provide sufficient oxygen for the microorganisms, and to provide for mixing in the aeration chamber and sludge holding tank for the operation of all air lift pumps. The second criterion requires more air and, therefore, determines the size of the blower. Without sufficient wastewater loading, continuous operation of the blower will lead to overaeration, which results in poor settling in the clarifier. Bleeding off excessive amounts of air or changing the motor sheaves sizes to adjust for the proper dissolved oxygen may result in insufficient mixing in the aeration chamber. A telephone call to the manufacturer or consultant can save a lot of needless problems. The best way to provide adequate mixing and aeration in a lightly loaded system is to operate the blower on a time clock (see next section).

An air filter should be attached to the intake of the blower. Typically it is also attached to a silencer. This filter may be of the disposable type (paper) or the permanent type (wire mesh), and must be kept clean for maximum blower life and optimum aeration. The advantage of paper filters is that they are disposable. Steel mesh filters must be washed periodically. Failure to keep the inlet air filter clean is one of the most common maintenance problems.

Typically, the blower provided will have a discharge pressure of 20 to 40 kPa (3 to 6 psi). A pressure relief valve should be installed on the blower discharge to protect it against excessive pressures, should an air line become plugged or an air valve be accidentally closed. This valve may be a dead-weight type as shown in Figure 5.4a or a spring-loaded type as shown in Figure 5.4b. The dead-weight type is more reliable, but must be kept well-lubricated. The gaskets on the spring-loaded type decay with time and must be replaced.

There must be a check valve on the blower discharge to help reduce the amount of water that enters the drop pipe when the blower shuts off. More importantly, because the two blowers use a common manifold, a check valve must be located near the discharge of each blower to prevent the blower that

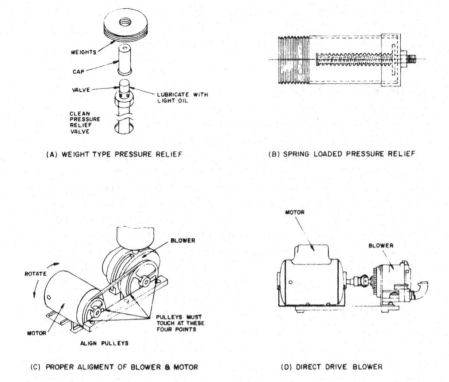

WEIGHTS
CAP
VALVE
LUBRICATE WITH LIGHT OIL
CLEAN PRESSURE RELIEF VALVE

(A) WEIGHT TYPE PRESSURE RELIEF

(B) SPRING LOADED PRESSURE RELIEF

BLOWER
ROTATE
MOTOR
ALIGN PULLEYS
PULLEYS MUST TOUCH AT THESE FOUR POINTS

(C) PROPER ALIGMENT OF BLOWER & MOTOR

MOTOR
BLOWER

(D) DIRECT DRIVE BLOWER

Figure 5.4 Typical alignment and air relief: (a) weight-type pressure relief, (b) spring-loaded pressure relief, (c) proper alignment of blower and motor, and (d) direct-drive blower.

is not in operation from rotating backwards and damaging both the blower and the motor.

If the rotation of the blower is reversed (for example, when a motor is replaced and it is wired backwards), a vacuum will be produced. Before connecting a new blower to the air distribution line, the blower should be operated for a few seconds to ensure that it is producing air at its discharge rather than drawing a vacuum. This procedure could prevent water from being sucked into the blower.

An electric motor is typically used to power the blower. The most common method of transferring power from the motor to the blower is by V-belts. The importance of the proper size sheaves on the motor and blower has already been discussed. Figure 5.4c illustrates the proper alignment of the blower and motor to prevent shaft bearing damage. Bearings may also be damaged by overtightening the V-belts. The operator should be able to depress the V-belt approximately 25 cm (1 in.) at the midpoint between the blower and motor for a properly tensioned belt. If two V-belts connect the blower and motor and one becomes damaged, both should be replaced. Dual V-belt systems take matched sets. The undamaged belt has stretched and worn with time and no longer matches the replacement belt even if it has the match number; this condition will make proper tension adjustment with the new belt impossible.

Figure 5.4d shows another method of connecting the blower and motor by using a direct drive. This method does not allow for adjustments to the blower's rpm as the V-belt and sheaves method does. For this arrangement, care must be taken to ensure the rpm of the motor match those required by the blower. Likewise, any replacement motor should also have the same rpm. Typically, the direct drive unit will not be found on anything except very small or very large systems.

MECHANICAL SURFACE AERATION. Oxygen transfer with mechanical surface aeration takes place primarily by either exposing small droplets of mixed liquor to the atmosphere, beating small bubbles into well-mixed water, or a combination of the two. The aeration takes place by the physical transfer of oxygen from the atmosphere into the droplet and by diffusion from the beaten bubbles into the mixed liquor. Just as the differences in diffused air center around diffuser type, the differences in mechanical surface aeration center around aerator blade type.

Figure 5.5 illustrates three typical types of aerators. One is a fixed drive with high-gear reduction between the motor and blade. The blade is on a vertical shaft and its rotation causes the mixed liquor to spray through the air. A similar type has the motor and blade mounted on a floating device, which keeps the blade or propeller below the surface. As the propeller turns, it pumps the mixed liquor up and out causing dispersal of the droplets and subsequent atmospheric transfer of oxygen. Another common floating aerator is the aspirator-type aerator. It is mounted at an angle into the mixed liquor, and a propeller drives the mixed liquor down and away from the device, as opposed to the floating aerator, while at the same time an aspirator tube draws air from above the surface into the turbulence of the pumped mixed liquor. This causes bubbles to be pumped into the aeration tank, and the bubbles transfer oxygen just as compressed-air diffused air does. All mechanical aerators have a high degree of energy input into aeration. As such, they are very dangerous pieces of equipment and utmost care must be taken when working with the reduction units. Have a lockout/tagout program in place, and follow it.

TIME CLOCKS. In the majority of package plants, it is not necessary to operate the aeration devices 24 hours per day. In fact, in many cases, that leads to the development of a biomass that is difficult to settle in the clarifier. During typical operation, time clocks that activate the aerator circuits several times each day are installed. Clocks are available that control alternating units and make it possible to have varying run times so that the operator can try to match incoming oxygen demanding loads. Figure 5.6 illustrates a typical time clock.

If the clarified effluent quality is poor, a change in run time is required, in conjunction with possible return sludge rate and/or wasting rate adjustment. The rule of thumb should be that the mixed liquor has a dissolved oxygen concentration of 2.0 mg/L within 10 minutes of the start of an aeration cycle.

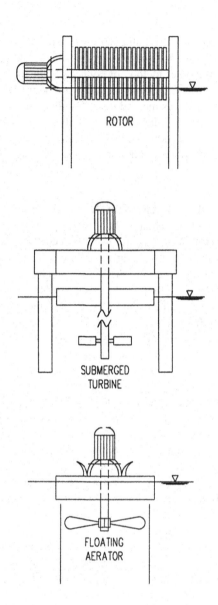

ROTOR

SUBMERGED
TURBINE

FLOATING
AERATOR

Figure 5.5   Types of mechanical surface aerators.

If the mixed liquor has become septic or anaerobic, it will not be possible to achieve 2.0 mg/L in 10 minutes. The aeration chamber of a plant receiving design flow should not be without air for more than a couple hours, depending on the degrees of treatment the operator is attempting to achieve. Figure 5.7 gives a typical aeration cycle for a plant that serves a campground that is heavily loaded on weekends and lightly loaded on weekdays. Aeration cycles could vary during the day (diurnal cycles) if most of the loading is received in a couple of relatively short periods of the day.

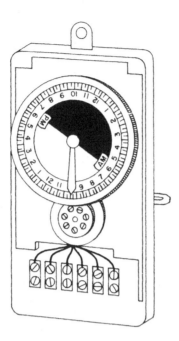

Figure 5.6 Typical time clock.

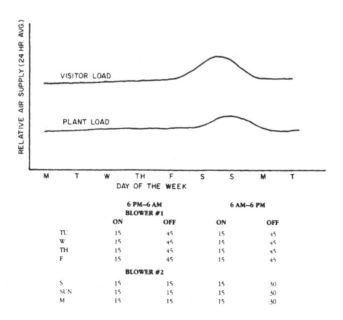

Figure 5.7 Typical aeration cycle.

	6 PM–6 AM		6 AM–6 PM	
	BLOWER #1			
	ON	OFF	ON	OFF
TU	15	45	15	45
W	15	45	15	45
TH	15	45	15	45
F	15	45	15	45
	BLOWER #2			
S	15	15	15	30
SUN	15	15	15	30
M	15	15	15	30

Adjustment of the return sludge rate will be necessary for a plant that operates blowers on time clocks. If the blower runs constantly, the return sludge rate can be less because the return sludge in the bottom of the clarifier is more dilute than the sludge in a clarifier with a timer. In the system with a timer, the sludge continues to concentrate during the off cycle and the thicker sludge must be lifted by the pumps when the blowers restart. Consequently, more air must be balanced to the return sludge pumps to prevent them from plugging.

RETURN ACTIVATED SLUDGE CAPABILITY

After the mixed liquor has been in the aeration chamber for 18 to 24 hours, it flows to the clarifier, where settling occurs. The settled biomass, which is now concentrated, is returned to the aeration chamber through the return sludge line. Return activated sludge (RAS) can be thought of as reinoculation of the biomass. The return sludge is made up of mixed liquor organisms that have continued the transfer, conversion, and flocculation process while in the clarifier, and as long as they had oxygen continued to eat. They are returned to the front of the aeration chamber because they are now hungry and will absorb the food coming into the plant in raw wastewater. The process of transfer, conversion, and flocculation continues. The RAS is typically discharged at the head end of the aeration chamber to provide maximum contact with incoming wastewater. Some plants may have two or more return sludge lines, and the flows from them must be balanced. Sludge returning from the clarifier has not been aerated for several hours; therefore, it will exert an oxygen demand in the aeration chamber. It is exerting the typical oxygen demand of microbes exposed to new food, and it is also exerting an oxygen demand of its own from some of the cell die-off or lysis that occurs in the clarifier.

Keeping the RAS running is extremely important. If incoming wastewater is not met by hungry microbes in the RAS, the biology shifts, and treatment is hindered (or non-existent). The constant resupply of hungry organisms is the key to activated sludge operating as a process (this is explained in greater detail in the section on the theory of activated sludge).

Knowledge of this need leads to what is often the biggest operating problem at a package plant. Focusing on the fact that the RAS must always be running, and considering how difficult it is to estimate flowrate from an air lift pump, operators all too often set the RAS rate way too high. This actually does not cause much of a problem in the aeration chamber; however, it plays havoc with the hydraulics in the clarifier. The shortened clarification periods result in solids washout from the clarifier. At this point, it begins to affect the operation of the aeration tank. Solids washout produces the same effect as sludge wasting; however, by its nature it is uncontrolled and will result in inadequate treatment (see Chapter 7 on sludge wasting for more information).

SKIMMER LINE

The other (smaller) line from the clarifier to the aeration chamber is the clarifier's surface skimmer. The skimmer returns floating solids from the clarifier to the aeration basin. It is critical that only the minimum air required be supplied to the skimmer. If more air is supplied than is needed, the diffusers or return sludge air lift pump will be robbed of air. Additionally, excessive wastewater will be returned to the aeration tank, thus decreasing the hydraulic detention time in the clarifier and inhibiting settling. The same result will occur as returning too much sludge. Vibration in the air lines may vibrate the packing nuts (on gate valves) for the air control valves loose, making air adjustment difficult. The operator should keep all valve packing nuts snug so that adjustments do not change between plant inspections. A better approach is to replace them with plug valves or corporation stops. It is generally not necessary to go to the extent of installing needle valves or another air control types of valve because the air adjustments will not need to be that precise.

FROTH SPRAY SYSTEM

The froth spray system consists of a submersible pump, located in the clarifier just below the surface, which pumps clear liquid from this chamber into a manifold located in the aeration chamber. The manifold is on the side opposite the air diffusers and has nozzles attached to allow the clear liquid spray to be directed down and to the sides. The froth spray reduces foam in the aeration chamber. The froth spray pump should be operated only when necessary. Figure 5.8 shows a typical froth spray nozzle.

Foam and scum may appear in the aeration chamber under certain operating conditions. The three most common occasions during which foam is produced are

(1) During initial startup (foam);
(2) When excessive detergents are present in the wastewater (foam); and
(3) When sludge remains in the treatment plant too long (biological scum).

The color of the foam during initial startup is white to very light tan in color and fluffy; it will dissipate as solids increase in the mixed liquor. Foam caused by detergents is extra light and may easily overflow the aeration chamber. The froth spray is especially effective in suppressing these two types of foam. It will be knocked down almost immediately. The biological scum caused by old sludge has a heavier consistency and will be much darker than foams. The best way to eliminate scum is typically to waste some of the old sludge out of the system. There are other conditions to consider in wasting, so review them

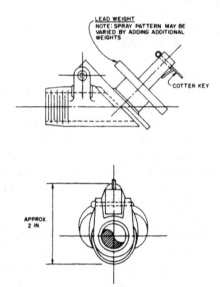

Figure 5.8   Typical froth spray nozzle.

before wasting to remove scum. Reasons and methods for wasting sludge are discussed in Chapter 7 on sludge wasting.

THEORY OF ACTIVATED SLUDGE

Secondary biological treatment is the conversion of organic materials in the wastewater to a biological cell mass (primarily bacteria) and the subsequent removal of this mass. The organic materials are used as food by bacteria. The food is converted into cell material as the microorganisms eat and reproduce. Once the food is in the form of bacterial cell, it is possible to remove it from the wastewater. This all takes place in the activated sludge system.

The following three conditions are necessary for the maintenance of any form of life: (1) living organisms (animal or vegetable), (2) acceptable food, and (3) suitable environment.

In the activated sludge process, the aeration tank provides the space for microorganisms to eat, grow, and reproduce. In this tank, microorganisms are supplied with food (incoming organic solids) and are provided with a suitable environment, proper pH range, dissolved oxygen, and a suitable temperature range to assimilate and convert the organic solids into activated sludge floc. Most municipal or domestic wastewater is within satisfactory pH and temperature ranges. The influent wastewater typically provides proper and sufficient food.

In most extended aeration plants, the wastewater temperature in the plant will be determined, to a great extent, by the surrounding atmospheric temperature (ambient temperature). The operation of the aeration chamber is affected in several ways by the temperature of the mixed liquor. During winter months, in cold weather areas, the activity of the microorganisms will be reduced. If the organic loading to the plant remains constant, more microorganisms will be needed for treatment during the winter than in the summer. The colder liquid also affects the dissolved oxygen concentration in the aeration chamber. The colder the liquid, the more oxygen it can hold in solution. As the liquid temperature increases, its ability to hold gases in solution decreases. For systems that are aerated intermittently with time clock control, the times and duration of the aeration cycle will be affected accordingly. The combination of activated sludge solids and incoming wastewater is termed *mixed liquor.*

ACTIVATED SLUDGE FORMATION. Activated sludge is formed in three distinct steps, as follows.

Transfer. When food comes into contact with microorganisms, the food is transferred to the organisms' cells through two actions: (1) adsorption to the cell wall and (2) absorption into the cell proper. To ensure proper transfer, the process requires that there be good mixing to provide contact between the bacteria and food and enough time to allow the transfer to take place. The rate of utilization of food by the organisms is controlled to maintain the following two conditions:

(1) Ensuring adequate time to assimilate the food is regulated by adjusting the sludge age, and
(2) The maintenance of an adequate quantity of organisms in the system is controlled by adjusting the food-to-microorganism ratio (F/M).

Conversion. The food matter (organic solids expressed as biochemical oxygen demand [BOD] or carbonaceous BOD [CBOD]) is converted to cell matter for growth and reproduction. The rate of growth is extremely important. If one does not consider this growth rate conversion step, the result will be an effluent which has not been properly treated. The conversion of food to cells is noticeable in the growth phase of the microorganism life cycle. Proper control of the F/M, cell residence time (CRT), and specific oxygen uptake rate (SOUR) will ensure that the conversion step is satisfied.

Flocculation. The final step in the formation of activated sludge occurs after the microorganisms have "picked up" the food and converted it to new cells. The organisms are now "full", their energy is dissipated, and their biological activity has slowed down. They attach to other organisms to form floc. As they enter a clarification tank to separate from the wastewater, they clump together into larger biological floc.

In the secondary clarifier, the activated sludge floc (containing the micro-organisms) separates from the liquid portion of the wastewater by gravity settling and is pumped back to the aeration tank as RAS to continue the process. The organisms again contact more food and dissolved oxygen, and the process cycle continues.

FOOD: FIVE-DAY BIOCHEMICAL OXYGEN DEMAND. The amount of oxygen required for the aerobic biological oxidation of the organic solids in wastewater is measured by the five-day BOD (BOD_5). The most commonly used test method is the five-day test at 20°C (68°F) method as detailed in *Standard Methods for the Examination of Water and Wastewater* (APHA et al., 1998).

The BOD_5 test was developed in an attempt to reflect the depletion of oxygen that would occur in a stream. Biochemical oxygen demand typically follows the pattern shown in Figure 5.9.

Oxygen uptake in the early phase of the reaction within the BOD bottle occurs mainly because of the oxidation of organic or "carbonaceous" compounds. Ammonia-nitrogen poses an analytical problem in measuring BOD_5. Nitrifying bacteria in raw domestic wastewater are insignificant in number and typically will not grow sufficiently during the five-day test period to cause a measurable oxygen demand. With the passage of time and removal of the carbonaceous organic matter, oxidation of ammonia occurs. The secondary effluent will contain a substantial number of nitrifying bacteria.

Because of the presence of the nitrifying bacteria, the BOD_5 test sometimes can be prone to error. To obtain a true measure of the treatment plant performance in removing organic or carbonaceous materials, the BOD_5 test requires inhibiting the nitrification. This is $CBOD_5$.

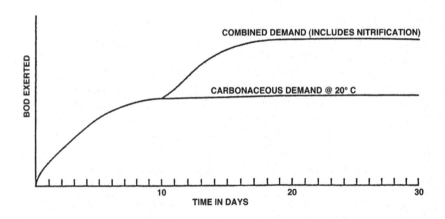

Figure 5.9   Biochemical oxygen demand exertion curve.

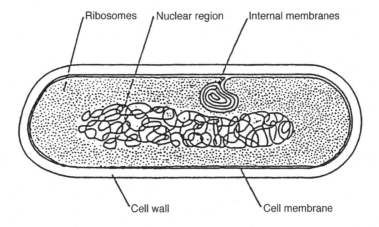

Ribosomes Nuclear region Internal membranes

Cell wall Cell membrane

Figure 5.10 Typical bacterium cell structure.

BACTERIA. Bacteria that use this food are living single-cell organisms. The bacteria, similar to humans, consume organic carbon as a source of energy for growth and reproduction. The bacteria, unlike the higher life forms typically observed under the microscope, are many times smaller and can only be seen with the use of special instruments. Figure 5.10 is a diagram of a typical cell (bacterium). The cell is comprised of the following.

- Cell wall. This provides the cell's shape and protection. If the cell membrane bursts or is ruptured, the bacterium dies.
- Cell membrane. This is a very thin structure, which allows the food and waste products to pass in and out of the intercellular membrane and then to the outside of the cell. This is the area that also stores food for later use.
- Internal membrane. This internal or intercellular membrane connects the cell membrane to the inside of the cell, and food and waste pass through it. It also serves as a cell organizer where many of the cell functions take place. There can be many of these within a single cell.
- Ribosomes. These are small particles comprised of protein and ribonucleic acid (RNA). These are typically attached to the intercellular membrane. In a single cell, there can be as many as 10 000 ribosomes. Synthesis of the food source to a protein takes place here. There may be several different types of ribosomes involved in the synthesis process.
- Nuclear region. The complete function of this is not exactly known. This is where the cell's chromosomes are found, and it exists as a long strand of deoxyribonucleic acid (DNA). The DNA is the key substance that determines the production of proteins, enzymes, and other cellular products. It also transmits this information to new cells during cell reproduction.

The process in which the bacterium uses the carbon source (food) is shown below:

$$\text{carbon} + \text{oxygen} + \text{bacteria} = \text{energy (heat)} +$$
$$\text{carbon dioxide } (CO_2) + \text{new cell growth} \qquad (1)$$

In the above reaction (eq 1), under ideal conditions, a single bacterium will reproduce or split into two individual bacterium approximately every 20 minutes and release 60% of the energy consumed in the reaction as heat.

REGULATION OF THE BACTERIUM. The actual process of converting organic matter to new cell mass, energy, and CO_2 is very complex. There can be several hundreds of steps or processes that take place before the waste can be stabilized or BOD removed. In some instances, a particular bacterium will break down the waste to a certain compound; then another will break it further, and so on. Each of the bacterium performing a particular function has a particular enzyme that can complete the work.

The process is controlled primarily by the particular bacterium's genetic capacity, which will ultimately determine which enzyme is produced and which waste it can synthesize or consume. The bacteria's environmental factors such as pH, temperature, waste concentration, availability of nutrients (ammonia [NH_3] and phosphorus [P]) and micro-nutrients (zinc [Zn], calcium [Ca], and magnesium [Mg], etc.) determine genetic capabilities. Ultimately, these factors determine if the bacteria will break down a particular waste or ignore it.

In the activated sludge process, there are hundreds of different types of bacteria; each breaks down a particular substance in the waste stream. A particular bacteria's level of predominance is determined by the environment, availability of food, and ability to store food. The method of storing food allows the bacteria to survive and maintain predominance when its particular food source is depleted.

When the food or waste characteristics change, so does the bacterial population. This is called *bacterial population dynamics*. Idealistically, if the operator has the same amount and type of food source for the bacterium and controls the environment at a steady state, the operator would maintain a balanced bacterial population. The result would be a very consistent output or effluent quality. As the influent waste characteristics change, the bacterial population changes, and subsequently there is a higher or lower growth phase for a particular bacteria. The bacterial population grows and declines with the presence or absence of a particular waste that it consumes. An example of this is shown in Figure 5.11. Note that the bacterial population growth is a logarithmic growth, given the proper environment.

The maximum breakthrough value (or discharge of a particular waste untreated) is a function of the following.

- Concentration and duration of influent spike (i.e., the total grams [pounds] applied over time);
- Initial concentration of the particular bacteria that can assimilate the waste;

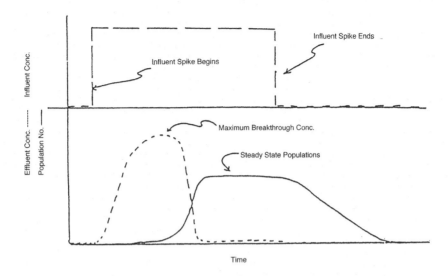

Figure 5.11   Dynamic response of a type of biological population.

- The growth rate of the particular bacteria or rate of substrate conversion; and
- Environmental conditions (dissolved oxygen, pH, nutrients, hydraulic retention time [HRT], etc.).

To minimize the breakthrough, the following can be done.

(1) Minimize potential tank short circuiting;
(2) Maximize HRT and/or sludge retention time (SRT) (if possible);
(3) Ensure proper environmental conditions (dissolved oxygen, pH, etc.).

NUTRIENTS AND MICRONUTRIENTS. Bacteria, like humans, need more than oxygen and a carbon source. Their main nutrient requirements are nitrogen and (ortho) phosphorus to sustain life. Bacteria use nitrogen in the form of NH in the production of amino acids, which form proteins. Phosphate in an inorganic form is converted to an organic form and is used in making polyphosphates, which are used in storing food for later use.

Micronutrients or trace elements are also necessary for bacteria, but in far less quantities than the nitrogen and phosphorous nutrients. A typical inorganic growth medium used for culturing bacteria is listed in Table 5.1. This simply illustrates the types of nutrients needed for bacterial life.

These trace elements should be present, at the concentrations noted, in the aeration tank to ensure proper bacterial growth.

BACTERIAL GROWTH. To understand how the food in the activated sludge process is used and how bacteria populations grow in response to the food, it is necessary to look at a growth curve.

Table 5.1 Typical inorganic growth medium.

Element	Concentration (mg/L)
Sodium	1.0
Potassium	3.0
Chloride	1.0
Magnesium	3.0 to 10.0
Calcium	3.0 to 5.0
Iron (Fe)	1.0 to 4.0
Manganese	0.02 to 0.05
Zinc	0.02 to 0.05
Molybdenum	0.02 to 0.05
Cobalt	0.02 to 0.05

ACTIVATED SLUDGE GROWTH CURVE

Figure 5.12 shows an activated sludge growth curve. If a large quantity of food is available in proportion to the quantity of microbes (the left side of the curve), the bacteria will use the food in two ways. First, they must obtain energy from the food to sustain basic life activities such as movement, heat production, and maintenance of internal and external structure. Second, the additional food

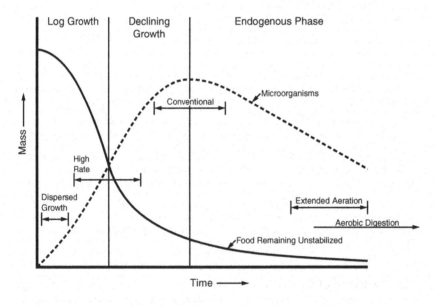

Figure 5.12 Activated sludge growth curve.

allows for production of energy used to produce new cells. When food is in excess, both of these processes occur simultaneously until the cells reach a maximum size. The cells then reproduce by dividing themselves and the process starts again. This rapid cell growth and reproduction is called *dispersed growth* and is dependent on the availability of large amounts of food. A high SOUR exists under this condition.

As the cells multiply, the food supply decreases. The rapid increase of microorganisms and the subsequent decrease in food supply is termed the *log-growth phase*. This phase has a very high F/M. Package plants do not operate in this phase because the lack of sufficient available microorganisms, for the large amount of food available, results in excess food being discharged into the effluent. The only times this condition could occur in the biomass is at startup and if the RAS pumping stops resulting in an aeration tank full of raw wastewater and no microbes. At this point, the WWTP is very susceptible to slugs of high-strength wastes.

As the food is used, the bacteria will soon outnumber the available food supply. Because of this, the microorganisms must work or compete harder for their food. As a result, they use the available food to support basic life functions. Therefore, a lesser amount of energy is available for reproduction and cellular growth. The reproduction cycle takes a longer period of time as a result. This slower rate of growth is called the *declining growth phase*. The bacteria are working hard for the available food, and this creates a relatively clean effluent.

If the biomass continues to progress in this manner, it will reach a stage at which food is so limited that growth rate and death rate become theoretically close to equal. The F/M is very low and starvation begins. The bacteria begin to break down the cellular materials or polysaccharides built when food was plentiful. This phase is called *endogenous respiration* or the *death phase*. This phase occurs with longer sludge ages and results in a much lower SOUR.

ACTIVATED SLUDGE PROCESS CONTROL

As mentioned at the beginning of this chapter, the activated sludge process will be addressed from two different angles. The first is rudimentary, or the basics and direct instructions for what is needed to get by. The purpose of this angle is to enable the operator, new to package plants, to get started and begin doing a good job of caring for the facility. The remaining angle is one that describes, in more depth, the factors that come into play in running a plant that are taken into consideration when problems arise. It is the knowledge of the second that makes the simplicity of the first possible.

After the general theory of the activated sludge process and the plant units necessary for operating the system become familiar, one must learn to control them. The purpose of the Activated Sludge Process Control section of this chapter is to help the operator learn how to control this process. By using test results and calculations to provide needed information on the activated sludge process, the operator will be able to make the necessary decisions to achieve the degree of treatment the plant is designed to provide.

The operator must make an overall evaluation of the plant and then decide on an operation strategy. The following factors must be considered.

DISSOLVED OXYGEN LEVEL IN THE AERATION TANK. It is necessary to maintain an aerobic environment for the microorganisms at all times. A range must be selected for the dissolved oxygen level in the aeration tanks. This range should be a minimum of 2 mg/L and never higher than 4 mg/L at the effluent point of the aeration tank. If the dissolved oxygen at the effluent port of the aeration tanks drops below 2 mg/L, the operator should provide more air to the system. If the dissolved oxygen gets as high as 4 mg/L, the amount of air is way in excess of what is needed and can be reduced. With this kind of control, the bacteria will always be living in an aerobic environment and proper mixing will be taking place in the tank. One way to look at dissolved oxygen is that it is excess or residual oxygen. The 2.0 mg/L ensures that there is oxygen available for microbe use. At 4 mg/L, energy is probably being wasted. The odor will indicate when there is too little oxygen. A sour odor will exist at conditions approaching septicity; however, if the odor is of rotten eggs or sulfur, septicity is occurring and the response is to put the aerators in hand and run them full time. The operator should also be checking for air leaks that may have robbed the process of its air.

RETURN ACTIVATED SLUDGE RATE. The RAS rate is an indicator of how fast and how much settled biosolid material is being removed from the bottom of the clarifier. The quality of settled sludge in different activated sludge modifications typically determines the starting point for RAS rate determinations. In conventional activated sludge processes, RAS rate may vary from 25 to 50% of the influent wastewater flow. Package plants use the extended aeration modification of the activated sludge process, and a 100% RAS rate is considered to be the benchmark rate. This means that if raw wastewater flow is 190 m³/d (50 000 gpd), RAS flow is also 190 m³/d (50 000 gpd). Because most package plants do not have the means to measure RAS rate, the operator must visually set flow to the proper rate. To find that rate, a simple bucket and stopwatch check may be made. Once the proper rate range is hit, the operator notes the intensity of flow out the RAS pipe. It is inexact and simple, but it is typically the best the package plant operator has on a day-to-day basis. Another simple rule of thumb is to set the RAS rate as low as it will go and still keep pumping over a period of days. This will typically result in the most consistent performance; at too low a rate, the pipe will plug. At too high a rate, the sludge blanket in the clarifier will raise and overflow.

WASTE ACTIVATED SLUDGE. Waste activated sludge is excess sludge that occurs from the continual growth of the microorganisms. Sludge must be wasted from the system to keep the F/M and sludge age in proper balance. The amount of sludge to be wasted can be calculated by using a sludge growth factor. Most well-operated package plants grow solids at a rate of 0.4 to 0.6 g (lb) of sludge for each gram (pound) of incoming BOD. A calculation using 0.5 g (lb) grown per gram (pound) BOD in the raw wastewater can be made to give the operator a starting point for wasting sludge out of the system. Fine-tuning of the waste rate is then made by maintaining the desired concentration of mixed liquor suspended solid (MLSS) in the aeration tanks.

The calculation is accomplished by multiplying the incoming flow in cubic meters per day by the milligrams per liter BOD to equal the grams per day load to the aeration tanks (mgd × 8.34 mg/L/mg BOD to equal the lb/day BOD load). The grams per day (pounds per day) of BOD are then multiplied by the growth factor (say 0.5) to find the grams (pounds) of sludge that must be wasted. The grams (pounds) of sludge to be wasted are then divided by the milligrams per liter concentration of waste sludge (actually the RAS) to give the amount to be wasted in cubic meters (lb ÷ mg/L × 8.34 = mgd).

$$\text{WAS flow (m}^3) = \frac{\text{Raw wastewater flow (m}^3) \times \text{mg/L BOD}_5 \times 0.5 \text{ lb grown}}{\text{WAS suspended solids (mg/L)}} \quad (2)$$

or

$$\text{WAS flow (mgd)} = \frac{\substack{\text{Raw wastewater flow (mgd) BOD}_5 \times \\ 8.34 \text{ lb} \times 0.5 \text{ lb grown}}}{\text{WAS suspended solids (mg/L)} \times 8.34} \quad (3)$$

It is important to keep in mind that WAS is taken directly from the RAS flow from the clarifier. It will probably be thicker than the aeration tank MLSS concentration, so care must be taken to use the correct concentration in the calculation.

Other calculations can be made to estimate the grams (pounds) of sludge to be wasted, using comparisons between MLSS inventory, sludge age, F/M, and MLSS (mg/L) concentration.

If the total grams (pounds) to waste figure is larger than normal (greater than 10 to 20% above normal) because of a correction in F/M or sudden decrease in loading, the additional wasting should be spread out over a period of at least a week. Rather than make a sharp change in the biomass, increase wasting by approximately 10% per day. If the requirement is to increase solids inventory, wasting may be decreased or, in certain drastic cases, stopped altogether. Excess solids are typically wasted by diverting part of the RAS to an aerated holding tank or aerobic digester.

Wasting should occur on as much of a schedule as is allowed for the package plant operator. Wasting a little bit on each visit to the plant will provide much more consistent results for the operator than having to waste large volumes monthly or even less frequently. Many times, plants are constructed in such a way that wasting is accomplished by turning off the air, letting the aeration tank settle, bringing in a tank truck, and pulling a load off the bottom of the aeration tank.

AERATION PERIOD. This calculation is necessary to tell the operator how long the wastewater and microorganisms are held in the aeration tanks. The aeration period is found by

$$\frac{\text{Capacity of aeration in m}^3 \times 24 \text{ h/d}}{\text{Influent flow, m}^3\text{/d}} = \text{Detention, h} \qquad (4)$$

or

$$\frac{\text{Capacity of aeration in gallons} \times 24 \text{ h/d}}{\text{Influent flow, gpd}} = \text{Detention, h} \qquad (5)$$

An extended aeration system is designed for a 24-h average aeration period.

INCOMING LOAD. The incoming load calculations inform the operator of the amount of the organic load going to the aeration system. This is the indicator of the strength of the waste to be treated. It is of utmost importance to note that it is the *quantity* of BOD that is important. Exceptionally high-strength wastes may have their particular operating problems, but control of the activated sludge process is based on the *quantity* of waste load. Two items must be known to determine proper operational controls.

Grams of BOD or suspended solids to the system:
g of BOD or SS = flow (m³) × mg/L (BOD or SS) (6)

or

Pounds of BOD or SS to the system:
lb of BOD or SS = flow (mgd) × mg/L (BOD or SS) × 8.34 lb/gal. (7)

This calculation is used to determine any loading (grams or pounds). Simply enter the appropriate concentration (mg/L) of the parameter in question and the appropriate flow.

Grams of BOD per cubic meter of aeration:
$$\frac{\text{g BOD(from above calculation)}}{\text{aeration capacity in m}^3} = \text{g BOD/m}^3 \qquad (8)$$

or

Pounds of BOD per 1000 cubic feet (cu ft.) of aeration:
$$\frac{\text{lb BOD (from above calculation)}}{\text{aeration capacity in cu ft/1000.}} = \text{lb BOD/cu ft.} \qquad (9)$$

MOVING AVERAGES. To buffer the effects of varying waste loads, a moving average is used. A moving average is nothing more than an average of the

---

most recent waste loads (the last seven or five values, etc.). It is expressed most commonly as a seven-day moving average (7 DMA), although a 5 DMA is occasionally used. A 7 DMA is calculated as discussed below.

When the latest figures (kg/d [lb/d]) (1075 on second calculation and 890 on third calculation) became available, the first and second day's values were dropped and the latest figure was averaged into the column. This provides a continuity of control. Moving averages smooth out the operator's reactions to the changes in the dynamics of the activated sludge process. Had the high 1075 kg (lb)/d loading figure been used by itself to determine an F/M, the result would have shown a serious shortage of mixed liquor solids inventory, and the operator's reaction would have been to cease wasting. By using the moving average (879 kg [lb]/day), the increase is accounted for, but the reaction to the calculation is moderate. By the time the third calculation data are in, the operator sees that the system has settled back down, and adjustments are minimal.

SLUDGE INVENTORY. Just as the quantity of incoming load is important, so is the total quantity of organisms available to treat the load. The term used to refer to the total biological mass (or biomass) is *sludge inventory*. It is expressed as MLSS inventory and as a volatile, or mixed liquor volatile suspended solids (MLVSS), inventory. It is determined as follows.

$$\text{Capacity of aeration tanks (m}^3\text{) on line* } \times \text{ mg/L MLSS or} \\ \text{MLVSS} = \text{g inventory (MLSS or MLVSS)} \tag{10}$$

or

$$\text{Capacity of aeration tanks (mil gal) online* } \times \text{ 8.34 lb/gal } \times \\ \text{mg/L MLSS or MLVSS} = \text{lb inventory (MLSS or MLVSS).} \tag{11}$$

*Some operators also consider the capacity of online clarifiers. The important thing is to pick a method of calculation and stick with it.

FOOD-TO-MICROORGANISM RATIO. The F/M tells the operator how many active microorganisms are in the system in relation to the food applied to it. By controlling the inventory, the operator controls the activity level and degree of treatment attained. The F/M is calculated by using the MLVSS inventory as an indication of active microorganisms. This gives a more accurate picture of the active mass in the aeration system.

The F/M is one of the most critical parameters for control of the activated sludge process. It must be calculated at least once each week based on a moving average (7 DMA) of influent load and more often when drastic changes occur in the influent load or sludge inventory. Because there is a limit to the amount of food that organisms can absorb, sufficient numbers must be available when the influent load fluctuates. Time is also critical to the organisms so they have sufficient time to assimilate and stabilize (absorb) their food. The F/M is calculated by dividing the grams (pounds) of BOD applied to the aeration tank by the grams (pounds) of MLVSS inventory.

$$F/M = \frac{\text{[g/d BOD applied]}}{\text{aeration capacity (m}^3) \times \text{mg/L ML VSS}} \quad (12)$$
$$\text{[g MLVSS inventory]}$$

or

$$F/M = \frac{\text{[lb BOD applied]}}{\text{aeration capacity (mil gal)} \times 8.34 \times \text{mg/L MLVSS}} \quad (13)$$
$$\text{[lb MLVSS inventory]}$$

The normal operating range is 0.05 to 0.2 for extended aeration. This means there are 0.05 to 0.2 g (lb) of food for each 1.0 g (lb) of microorganisms. The higher the F/M, the higher the sludge growth rate and activity level of the microorganisms. Conversely, the lower the ratio, the lower the growth rate and activity level. This rate of growth is also affected by temperature, and it may be necessary to maintain a lower F/M in winter than in summer.

There is generally very little the operator can do to control the BOD applied, so the F/M is controlled by controlling the inventory of solids. For example, if experience has indicated that an F/M of 0.10 g/g (lb/lb) is desirable and the influent load averages 260 g (lb)/day BOD, calculate as follows.

(known)   (unknown)

$$\frac{260 \text{ g (lb)/day BOD}}{0.10 \text{ g/g (lb/lb) F/M desired}} = 2600 \text{ g (lb) MLVSS required} \quad (14)$$

Then 2600 g (lb) MLVSS is the inventory required. Wasting can be increased or decreased to bring the inventory into the proper range.

The terms *old sludge* and *young sludge* also relate to the F/M. The higher the ratio, the younger and more active the sludge.

SLUDGE AGE OR CELL RESIDENCE TIME. The sludge age or CRT (the terms *sludge age* and *CRT* are used interchangeably in this chapter) is an indication of the amount of time the activated sludge is held in the aeration system. Because time is needed by the bacteria to assimilate the food and stabilize it before they are returned back to or wasted from the aeration system, time in the aeration system is important. Sludge age should be considered as a tool for controlling the time required for microorganisms to assimilate the available food (BOD). Sludge age is a measurement of the time required for the existing biomass to be completely removed from the system through proper wasting procedures. It is simply how much time the activated sludge solid remains in the system. Sludge age or CRT is calculated as follows.

$$\frac{\text{total g (lb) of MLSS inventory}}{\text{total g (lb) of solids removed per day}} = \text{CRT (days)} \quad (15)$$

Cell residence time has also been termed *solids (sludge) retention time* or *mean cell residence time*. Minor variations exist in the calculations of sludge age, CRT, and SRT. Some operators use MLVSS. Some consider the effluent suspended solids as part of the solids removed each day. The important thing is that the operator chooses the control to be used and remains consistent in its use. Cell or sludge residence time, sludge age, and F/M are all controlled by the WAS system. When solids are wasted to maintain these controls, microorganisms are kept at a relatively constant level of activity. Maintaining a proper level of activity ensures the operator of predictable reductions of BOD through the aeration system.

The target sludge age should be selected on the basis of the wastes coming into the plant. If the plant is subject to frequent organic shock loads or is required to nitrify, the sludge age should be 20 to 30 days. If the plant receives a normal wastewater flow and load, a shorter sludge age of 15 to 20 days is typically sufficient. The sludge age is important because it indicates the time the organisms are kept in the system and also gives an indication of what the general energy or activity level should be. It is very helpful for most plants to operate on a short sludge age because the activity level of the microorganisms is higher. It does not require many solids to be stored in the activated sludge system. The shorter sludge age thus will be less susceptible to washout of solids in the effluent if the plant is subject to high storm flows. Longer sludge ages are required for wastes that are difficult to break down and also for systems that are designed to nitrify.

Sludge age should be calculated at least once each week based on the average wasting and current MLSS inventory. Using seven-day average results for wasting tends to neutralize the variations obtained from the test results.

SLUDGE VOLUME INDEX. One of the most important goals of a well-operated package plant is to produce a floc which will separate well in the final clarifiers. Overtreated or undertreated wastewater will not produce a good settling floc. A floc which settles well will carry fine suspended solids to the bottom of the clarifier with it. The laboratory test and calculation which best indicates the settleable quality of the activated sludge floc is the sludge volume index (SVI). It is an index number that indicates the quality of the sludge flowing to the clarifier. The ideal SVI is approximately 100. If the SVI is higher than 150, it is an indication that the floc has poorer settling qualities and that the treatment strategy may need adjustment. Rechecking the F/M and sludge age will typically provide some answers. If the problem still exists, then more serious conditions are indicated and checking of the load, microorganisms, and other parameters must continue. Do not be too quick to make rapid adjustments. Typically, the higher the SVI, the clearer the effluent, and many plants can operate well with a high SVI.

The SVI is the volume (in milliliters) occupied by 1 gram of sludge (dry weight) after 30 minutes settling. It is calculated as follows.

$$\text{SVI} = \frac{1000 \times \text{mL settled in 30 min in a 1-L cylinder}}{\text{mg/L MLSS}} \qquad (16)$$

It is very important for the operator to understand the unique qualities of the plant's individual sludge. For example, sludge with an SVI of 200 mL/g typically produces crystal-clear effluent, but may be difficult to handle in other portions of the WWTP. Nitrified sludge does not often produce an SVI of 100 in a package plant, but accomplishes the jobs intended (i.e., nitrification and clarification) very well. High SVI sludge is much more susceptible to storm flow washout.

Degree of nitrification required, operating capabilities of the secondary clarifier, and reactions to storm flows are all factors to be considered when determining whether or not a specific SVI is trending towards trouble. If the SVI changes, all other operating parameters should be investigated for changes and the total picture looked at. All the controls are interrelated and each affects the other.

SPECIFIC OXYGEN UPTAKE RATE. The level of activity of the microorganisms in the activated sludge system is indicated by its SOUR. As with any living thing, the harder it works, the more oxygen it needs to do that work. Operating experience will indicate at which rate a specific system works best. Most extended aeration activated sludge plants operate within a range of 6 to 10 mL/h/g volatile suspended solids (VSS). Specific oxygen uptake rate can be a rapid indicator of impending problems. A very low SOUR indicates a lack of available food for the organisms present (F/M too low) or possibly a toxic shock load that has killed a large portion of the biomass. A high SOUR indicates an organic overload, which could be caused by (1) improper operation of upstream treatment units, (2) excessive wasting procedures, (3) storm flows flushing settled organics from the sewers, or that the RAS pump has been inoperable (although that should have been visually apparent first). All of these causes make the F/M too high.

When properly used, the SOUR can be one of the most valuable control tools in the WWTP, because it can be rapidly obtained and is directly related to the actual activity of the microorganisms. To be used to its fullest, it must be viewed in relation to all the control tools previously discussed.

The procedure for determining SOUR is discussed in *Standard Methods* (APHA et al., 1998).

MICROSCOPIC EXAMINATION OF ACTIVATED SLUDGE

One need not be a microbiologist to effectively use a microscope in a WWTP. At its most basic level, microscopic control is a matter of being able to identify five to six general types of microorganisms, count them (look at the numbers in relationship to each other), and interpret what the predominances mean. Relative predominance, or "how many of these are there in relation to those,"

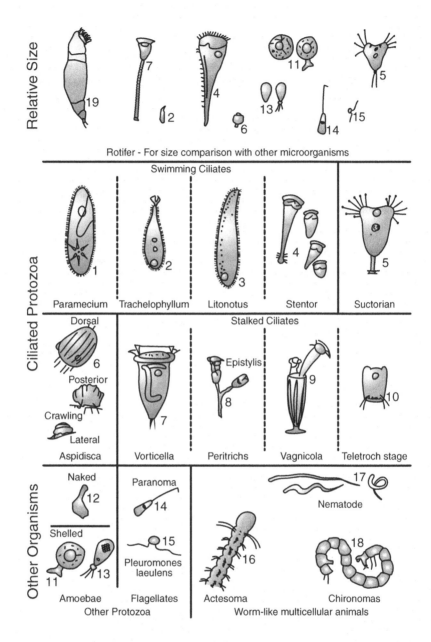

Relative Size

Rotifer - For size comparison with other microorganisms

Ciliated Protozoa

Swimming Ciliates

Paramecium   Trachelophyllum   Litonotus   Stentor   Suctorian

Stalked Ciliates

Dorsal
Posterior
Crawling
Lateral
Aspidisca   Vorticella   Peritrichs   Vagnicola   Teletroch stage

Epistylis

Other Organisms

Naked
Shelled
Amoebae   Flagellates   Actesoma   Chironomas
Other Protozoa   Worm-like multicellular animals

Paranoma
Pleuromones laeulens
Nematode

Figure 5.13   Common types of protozoa and metazoa.

indicates the general character of the entire biomass, or sludge quality. Figure 5.13 shows some common types of microorganisms found in activated sludge. The very general types mentioned previously are

(1) Amoeboids (11, 12, and 13). Small organisms with changing shapes found primarily in very young sludge.

The Aeration Chamber   61

(2) Flagellates (14 and 15). Very mobile organisms that have one or two tails that whip about rapidly. They are typically predominant in young sludge.

(3) Free swimming ciliates (1, 2, 3, 4, and 6 [Number 6 is a crawler, and for convenience, is classified a free swimmer]). These are in predominance in the younger range of good sludge quality. Cilia are hair like appendages surrounding their bodies, and cilia movement is what causes the organism to move about in its search for bacteria to eat.

(4) Stalked ciliates (5, 7, 8, and 9). These are in predominance with free swimmers, and they indicate a good sludge quality. They are, for the most part, stationary, i.e., attached to a floc particle by a stalk.

(5) Rotifer (19). These are the most easily identified organism in activated sludge. Rotifers will be present in good settling sludge and will be predominant in older sludges.

(6) Worms (16, 17, and 18). These are typically the largest organisms that will be seen under the microscope. They occur far up the food chain, indicating an old sludge, and may signal impending problems in a system.

Figures 5.14 and 5.15 show the significance of relative numbers of organisms in relationship to sludge quality. In Figure 5.14, the two columns under "Stragglers" are typical of very young to youngish sludge. The center columns under "Good Settling" are typical of conventional activated sludge. The two columns under "Pin Floc" indicate older to very old sludge. With the exception of the two extremes, there is no cut-and-dried good or bad. Experience will teach which predominance is best for a particular WWTP. Most package plants are designed to and will operate to the right or older character of sludge. If a

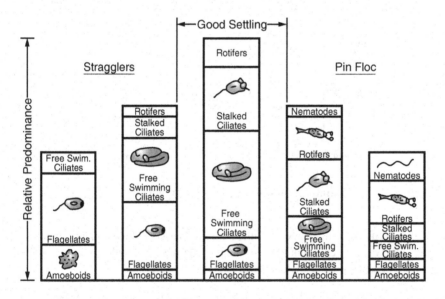

Figure 5.14 Relative number of microorganisms versus sludge quality.

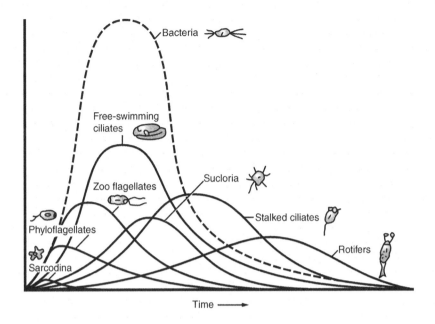

**Figure 5.15   Growth and predominance curve for common activated
sludge organisms.**

particular system must nitrify, or if shock loads indicate that it should carry an
older sludge, one may expect to see stalked ciliates in predominance with
lesser numbers of flagellates and free swimmers.

As with all the other controls discussed, use of the microscope should be
viewed in conjunction with information obtained from other tests.

FILAMENTOUS BACTERIA. Bacteria are the workers of the biological
treatment system. It is bacteria that use the organic materials as a food source
and thereby purify raw wastewater. There are many different types of bacteria
in activated sludge, and specific identification typically serves little purpose to
the WWTP operator. In activated sludge process control, it is sufficient to view
bacteria as falling into the following two general groups.

(1) Floc formers are, for the most part, the "good guys". They clump
     together into well-defined distinct and dense floc particles that separate
     from the mixed liquor and settle well.
(2) Filamentous bacteria are characterized by their hair and branch like
     appearances. They are strands of bacteria that form chains; this
     physical structure prevents them from forming a dense floc particle.

Filamentous bacteria are essential to the formation of a healthy activated
sludge. The filament forms a "backbone" or "skeleton" to which floc formers
attach. When floc settles in a clarifier, the well-formed floc, with its filamentous

---

The Aeration Chamber   63 backbone, provides a filtering action. Smaller solids that do not settle well are trapped or attached to the well-formed floc. The end result is a clearer effluent.

Problems arise when the filamentous strands extend too far from the floc body. The filament strands begin to hold floc formers apart, rather than aid in floc formation. This reduces the overall density of the sludge and the result is hindered settling and poor compaction once settled. The condition is called *bulking*. Basically, it describes an activated sludge that takes up more space than it should; it has become bulky. Should the overabundance of filaments continue, the result is almost always solids washout from the clarifier.

Scum on aeration tanks can often be caused by a type of filamentous bacteria called *Nocardia*. Under the microscope, it has a branch like appearance. *Nocardia* typically occurs in a sludge that is too old. Because package plants, by design, operate on the old side of the growth curve, *Nocardia* scum is an ever present danger. There will most likely always be some. It is overabundance that causes problems.

Filamentous predominance may be caused by a number of conditions, but most instances can be traced to one or two of the following: (1) too low an F/M, (2) too low a dissolved oxygen level, (3) septic wastewater/sulfide presence, (4) too low a pH, or (5) nutrient deficiency (nitrogen and/or phosphorus are the most common). The proper correction of a filamentous problem is to cure the real cause. Temporary actions such as chemical dosages or chlorine or hydrogen peroxide serve only to buy the operator time. The cause of the problem must be identified and corrected, or the filament predominance will continue or return.

Several different types of filamentous bacteria are associated with the five conditions previously discussed. Table 5.2, reprinted from the *Manual on the Causes and Control of Activated Sludge Bulking and Foaming* (Jenkins et al., 1993), details the types of filaments most often observed for the conditions listed.

The filaments should be identified using a microscope equipped with a 10× lens and 100× oil immersion lens. A wet-mount slide should be prepared and viewed using the 10× lens to determine the general abundance level of the filaments present. Then, the slide should be viewed using the 100× oil immersion lens to determine the specific filaments present. Some filaments

Table 5.2   Dominant filament types as indicators of conditions causing activated sludge bulking.

Suggested causative conditions	Indicative filament types
Low dissolved oxygen	Type 1701, *S. natans*, *H. hydrossis*.
Low F/M	*M. parvicella*, *H. hydrossis*, *Nocardia* sp., types 021N, 0041, 0675, 0092, 0581, 0961, 0.803.
Septic wastewater/sulfide	*Thiothrix* sp., *Beggiatoa* and type 021N.
Nutrient deficiency	*Thiothrix* sp., *S. natans* type 021N, and possibly *H. hydrossis* and types 0041 and 0675.
Low pH	Fungi.

Table 5.3   Filamentous organisms identification.

Organism	Description
Type 1701	Round ended rods, gram−, N−, attached growth, sheath.
Type 021N	G−, N−, beer barrels, big filament, sometimes sulfur and gonidia (rare, but sometimes see a foot) pinched.
Type 0041	Big filament, rectangular cells, G variable, N+ with dots, sometimes N attached sheath, attached growth.
Type 0675	Small 0041, rectangular cells, G variable attached growth.
Thiothrix (2 types)	G−, N−, sulfur granular, rectangular cells, cannot have rosettes or gonidia. No. 1 is big. No. 2 is small
Sphaerotilus natans	G−, N−, round ended rods, false branch, no attached growth, has sheath, has 2 or 3 granules in each cell.
H. hydrossis	G−, N−, thin filament, looks like needle, has sheath (must use 1000× to see).
Microthrix parvicella	G+, N+ granules, thin <1/u, coiled.
Nostocoida limicola	Looks like blood corpuscle, dark, can be both G±, N±, looks like stalking ladder, precoiled 3 types—typically group as a whole.
Type 1851	Attached growth, G variable, N−, bundles, attached growth at 90 deg angles.
Fungus	Big filament, true branches.
Type 1863	G−, N− free floating.
Nocardia	Tree antler branch G+, N+.

are very difficult to identify unless stained. The two types of staining methods most commonly used are the Gram and Neisser stains. Table 5.3 lists the staining and other general characteristics to be used to aid in filament identification (Note: Gram positive will appear blue or purple; Gram negative will appear red; Neisser positive will appear blue or purple; Neisser negative will appear yellow).

The *Manual on the Causes and Control of Activated Sludge Bulking and Foaming* (Jenkins et al., 1993) contains additional details and photographs of the filaments most commonly observed in activated sludge. Because of the susceptibility for package plants to be overcome by filamentous overgrowth, the manual should be part of every operator's reference library.

AERATION TANK FOAM. The formation of foam or scum on the aeration tank surface can be caused mainly by one of three items listed below.

(1) Young sludge. This is a white, frothy-type foam that is typically observed when the plant is operated at a young sludge age, during a process startup, or with an extremely high F/M. This will occur when the WWTP is extremely overloaded and the operator will see a high SOUR at this time. Young sludge typically dissipates when the biomass

volume increases and adapts to the current waste load, or the slug load subsides and the biomass assimilates the slug. This type of foam can typically be controlled easily by water spray or anitfoam chemicals.

(2) Chemically induced. This is from the presence of chemicals which (1) are difficult to biodegrade, (2) create a high surface tension, or (3) contain a surfactant that by its characteristics creates a foam. To reduce or eliminate this type of foam, the bacteria must quickly break it down or eliminate the chemical from the wastewater process. This type of foam can be controlled only to a limited degree by a water spray or antifoam chemicals.

(3) Nutrient control. When nutrients and/or micronutrients are deficient, the bacteria themselves can create foam. This is because the bacteria form a surface-active extracellular material that, when agitated, creates a foam. This is why it is important that the nutrients be controlled.

BASIC APPROACH TO PACKAGE PLANT MICROLIFE. The majority of package plant operators do not have access to microscopes and staining materials, although that situation is improving over time. So, how does the everyday operator take into account the microbiology of the package plant? Knowledge of what is occurring at the microscopic level points the operator in the right direction when making decisions. For example, a plant that had been running well and maintaining aerobic conditions is suddenly close to but not quite septic. The blowers are running and the timer is set at the same settings. Common sense indicates that the tank is not getting enough air so the operator looks for an air leak or plug or valve that has vibrated open (most likely on a skimmer). Knowledge of plant microbes indicates that there are only two ways for the oxygen to be too low. Either not enough is being put into the tank, or the demand by the microbes is higher than normal. There are no leaks, so something is demanding unusual amounts of oxygen (and thereby work) for the microbes. An overload is likely. The operator's response should be to increase attention at this plant to as frequently as twice a day, because it had been in good operation before this. If the biomass cannot work its way out of the problem, aeration must be adjusted.

If a problem of poor settleability persists and the operator is satisfied that it is not caused by too high a rate of return from RAS or the skimmer, knowledge of the plant's microlife suggests that the operator seek professional assistance. This could be from a local large municipal facility where the operators probably have the tools to help diagnose the problem, from a consultant, or from the authors of reference texts.

NUTRIENT CONTROL

Typically, package plants are designed to control one nutrient: ammonia-nitrogen (NH_3-N). This nutrient can cause excessive growths of rooted and/or floating aquatic plants in the receiving waters. These plants may (1) cause fluctuations in the dissolved oxygen level of the receiving water and thus discourage the growth of fish, (2) clog a stream or river, making it undesirable for recreational use and increase chances of flooding, (3) cause taste and odor problems in water supplies, and (4) die and decay, causing a continuing load on the oxygen resources of the stream.

NITRIFICATION. Nitrogen is commonly found in the environment in its various forms (organic [org-N], ammonia [NH_3-N], nitrite [NO_2-N], and nitrate [NO_3-N] nitrogen). Wastewater typically contains appreciable amounts of nitrogen. Total nitrogen concentration in domestic wastewater typically ranges from 15 to 50 mg/L, primarily in the organic and ammonia forms.

Ammonia-nitrogen is undesirable and potentially harmful when discharged to receiving streams. When present in sufficient concentration and at certain pH values, it is toxic to fish. Bacteria in the WWTP will oxidize NH_3 to the more stable form of NO_3-N, thus reducing significantly the depletion of oxygen and ammonia toxicity in the receiving stream. This oxidation by bacteria is called *nitrification*.

Nitrification in the extended aeration package plant is accomplished in the aeration tanks. Using long aeration periods and properly controlled sludge ages, the extended aeration process can effectively accomplish nitrification.

There are a number of items that will have an effect on the nitrification process, including

(1) Temperature. As the liquid temperature in the aeration system decreases, the bacterial growth rate is greatly reduced. It is, therefore, much more difficult to maintain nitrification under winter operating conditions. Nitrifying organisms are, by nature, slow growing and are very sensitive to changes in their environment. Greater MLSS concentrations can increase nitrification during colder periods.

(2) Ammonia-nitrogen concentration. If the influent NH_3-N concentration is very low, it will limit the growth rate because the nitrifying bacteria depend on NH_3-N as their energy source.

(3) Dissolved oxygen. Low dissolved oxygen levels will inhibit the growth of the nitrifying bacteria. The conversion of NH_3-N to NO_3-N requires a high amount of oxygen. The operator should strive to maintain 2 to 4 mg/L of dissolved oxygen in the mixed liquor. This range should be maintained throughout the entire aeration tank.

(4) pH. As pH decreases below 7.0, nitrification is inhibited. The nitrification process itself lowers the pH (7.2 kg of total alkalinity is consumed per kilogram of NH_3-N converted to NO_3-N). Optimum pH for nitrification is 8.4 at 20°C (68°F); however, in the plant it typically must occur in the practical range of 7.4 to 7.8. This is dependent on time, temperature of aeration, and influent alkalinity.

(5) Toxic metals and organic compounds. Some heavy metals and organic compounds can inhibit nitrification severely. As stated previously, the

nitrifiers are very sensitive to changes in their environment. Control of industrial waste discharges is mandatory.

Both temperature and pH significantly affect how much time is required for the bacteria to remove the NH_3-N. In warmer weather, it is easier to maintain nitrification.

Alkalinity is also important to maintain proper nitrification. During the conversion of ammonia to nitrate (NO_3), mineral acidity is produced. If sufficient alkalinity is not present, the pH of the aeration system will drop and nitrification will be inhibited. In systems where the drinking water has a historically low alkalinity value, which may stress the nitrification process, it may become necessary to add alkalinity to complete nitrification.

In the activated sludge process, the degree of nitrification depends on the sludge age. Process controls may indicate the need to increase sludge age to achieve desired results. The longer sludge age that is characteristic of extended aeration lends itself naturally to the nitrification process. A longer sludge age prevents nitrifying organisms from being lost from the system when carbonaceous wasting occurs and permits the buildup of an adequate population of nitrifiers. When compared to the normal bacteria in the activated sludge, the nitrifying bacteria have a very slow growth cycle.

NITRIFICATION PROCESS THEORY. Nitrification is the biological conversion of nitrogen in the form of ammonia (NH_3) to nitrogen in the form of NO_3. Nitrification is accomplished by providing the amount of oxygen required (4.6 kg of oxygen per kilogram of NH_3-N converted) and the proper biomass to complete the conversion.

Nitrogen enters the wastewater plant mostly in the form of NH_3. It is converted to nitrite (NO_2^-) when consumed by *Nitrosomonas* bacteria that exist in the aeration tank. The NO_2^- is then used by *Nitrobacter* bacteria in their metabolic process, and the NO_2^- is converted to NO_3. This conversion of NH_3 to NO_3 is called *oxidation* of the nitrogen.

d: superscript ? (2X)

In the presence of oxygen (O_2), CO_2, and bicarbonate ion (HCO_{3-}), NH_3 is also used to help synthesize additional *Nitrosomonas* and *Nitrobacter* bacteria. The HCO_{3-} is a natural substance found in most waters and greatly contributes to the amount of alkalinity in the water. This alkalinity is consumed in this synthesis of new bacteria and because it is necessary for the reaction, it must be present in the wastewater. Alkalinity (measured as calcium carbonate [$CaCO_3$]) is consumed at a rate of 7.2 kg for every kilogram of NH_3-N that is consumed.

Cell yield is the term describing how much new cell synthesis will occur per given amount of food that is added. Ammonia is the food in this case. It can be expected that 0.04 to 0.29 kg VSS will be formed per kilogram NH_3-N for *Nitrosomonas*, and from 0.02 to 0.084 kg VSS formed per kilogram NH_3-N for *Nitrobacter*.

DENITRIFICATION. *Denitrification* is the biological conversion of nitrates to nitrogen gas (N_2) by bacterial metabolism. Denitrification can only be accomplished under anoxic or anaerobic conditions (0.0 to 0.5 mg/L dissolved oxygen levels). In a quiescent environment, such as a clarifier or lagoon, denitrification can be observed by the bubbles of N_2 breaking from the liquid surface. A contributing factor to denitrification is the availability of a carbon source; typically the carbon source is CBOD.

In an environment devoid of oxygen, CBOD-removing organisms can breakdown the NO_2 and NO_3 and use the oxygen from them for respiration. The unused nitrogen is released as a gas. The organisms obtain oxygen from the NO_2 and NO_3 only if there is not enough actual dissolved oxygen available. In the clarifier, low oxygen concentrations exist in the sludge blanket. Denitrification also occurs in the aeration tank during the "blower-off" cycle. The microbes continue to use all available oxygen, but when they have settled into a blanket at the bottom of the aeration tank, they quickly consume all that is available and denitrification can proceed. In systems where denitrification is designed into the package plant, some mixing mechanism (in the absence of aeration) will most likely be installed. The following parameters affect the denitrification process.

(1) Temperature. Decreasing temperature of the liquid slows the activity of the microorganism and decreases capacity to denitrify.

(2) pH and alkalinity. Denitrification has been shown to proceed most efficiently at a pH level range of 7.0 to 7.5, and is reduced substantially when the pH drops below 6.0 or climbs above 8.0. Alkalinity consumed during nitrification is released during denitrification. The amount released is a little less than one-half the amount of alkalinity required for nitrification. Therefore, the overall alkalinity balance in the system shows a need of incoming alkalinity for nitrification.

(3) Organic substrate. Different organisms are responsible for denitrification than for nitrification. Heterotrophic organisms, those that consume organic compounds for energy and carbon source for synthesis, are required for denitrification. Autotrophic organisms, those that use CO_2 as a carbon source and inorganic compounds (such as nitrogen compounds) for energy, are required for nitrification. Autotrophs can consume materials within their own organism to survive. There must be a carbon source in the form of an organic compound for denitrification to occur.

(4) Oxygen availability. If oxygen (O_2) is present for use by aerobic heterotrophic organisms, the breakdown of NO_2 and NO_3 will reduce significantly. When oxygen concentration is low enough so that the NO_3 is used as an O_2 source, the total oxygen requirement for CBOD reduction is decreased. It is estimated that 2.6 kg of oxygen is supplied for every kilogram of NO_3 that is reduced.

(5) Toxic metals and organic compounds. Some heavy metals and organic compounds can inhibit nitrification severely. As stated previously, the nitrifiers are very sensitive to changes in their environment. Control of industrial waste discharges is mandatory.

SUMMARY

The aeration tank is the workhorse of the package plant. It is in this tank that the process of treatment occurs. The operator must use eyes, ears, and nose in addition to all that he or she has learned to maintain control of the system. In many cases, the operator's senses will alert to problems at the plant before they are discovered (the feeling that something is just not right).

Note the appearance of all the items at the plant. Over time, the operator builds a mental database of how things should look. The aeration tank contents will be a certain color or shade of brown. If it changes, there was a reason, and it demands investigation. It may turn out to be inconsequential, but it indicates change. Increased scum formation may well be signaling that it is time to waste sludge from the mixed liquor system. Return activated sludge and skimmer discharge rates will tell the operator a lot about what to expect at the clarifier. Discharges that are just blowing out at high rates and the lack of discharge alarms indicate that problems are happening in the clarifier and the effluent. The rolling pattern of the air diffusers can indicate to the operator that plugging is occurring or that a diffuser or lateral may have broken. A boiling action at or near a diffuser drop indicates that too much air is coming up at one location. The boiling action indicates that the air is not being diffused and is indicative of a break of some kind. The opposite situation, or lack of water activity, may well indicate an area experiencing plugging. It will not take the operator long to get a feel for the required mixed liquor concentration. Visually, the operator can tell whether the mixed liquor is thick or thin. This does not pass for laboratory testing, but does provide the operator an "eyeball" on the operation. Observations in a settleability test, depth to blanket, and clarity in the clarifier or scum floating on its surface all tell the operator how well treatment in the aeration tank is preceding and whether adjustments are necessary.

Sounds can alert the operator before he or she even arrives at the plant site. Unusual sounds emanating from blowers or pumps foretell problems and allow the operator to correct the situation without a breakdown and its subsequent odors and complaints. Common sounds include the mechanical whining or clanking that accompanies a blower beginning to fail, squealing of belts improperly adjusted or preparing to fail, hissing of air escaping from faulty flexible links, or the sucking sound that accompanies a pump that has run dry from a faulty control switch not stopping it.

The package plant will develop its own characteristic set of odors, and the operator will quickly be able to recognize them. Some will be normal and the operator should anticipate them. Unusual odors may well signal problems; for example, raw wastewater will have a musty yet not terribly unpleasant odor. Septic wastewater is offensive and has the rotten egg odor of hydrogen sulfide. The importance to the operator is why the wastewater went septic and whether there are more problems lurking upsewer. Another telltale odor to the operator is the sweetish sour smell of a mixed liquor that has been inadequately aerated. This is probably the most valuable odor tool to the operator because it alerts to a problem just before it becomes one. Unfortunately, it is difficult to describe the various odors, but fortunately they are characteristic enough that, once experienced, the operator recognizes them easily.

It is essential that the operator understand how inseparable the aeration tank and clarifier are. Biological treatment can and will occur if bacteria and food sources are put together with some oxygen. This does not become a treatment plant until the treated wastewater can be separated from the biomass. How well that separation occurs depends on what occurs in the aeration tank, and what happens in the clarifier (e.g., RAS or WAS) determines what will happen in aeration. How the two units work together determines the compliance status of the package plant.

REFERENCES

American Public Health Association; American Water Works Association; Water Environment Federation (1998) *Standard Methods for the Examination of Water and Wastewater*, 20th ed.; Washington, D. C.

Jenkins, D., Richard, M., and Daigger, G. (1993) *Manual on the Causes and Control of Activated Sludge Bulking and Foaming*. Lewis Publishers: Lafayette, California.

SUGGESTED READINGS

U.S. Environmental Protection Agency (1977) Process Control Manual for Aerobic Biological Wastewater Treatment Facilities. EPA 430/9-77-006, Washington, D.C.

U.S. Environmental Protection Agency (1978) Field Manual for Performance Evaluation and Troubleshooting at Municipal Wastewater Treatment Facilities. EPA 68-01-4418, Washington, D.C.

Water Environment Federation (1994) Basic Activated Sludge Process Control—PROBE Series. Alexandria, Virginia.

Water Environment Federation (2002) Activated Sludge. Manual of Practice OM 9, Alexandria, Virginia.

Chapter 6
Clarifiers

GENERAL DISCUSSION

The basic types, design criteria, operation procedures, and troubleshooting of final clarifiers are discussed in this chapter. Basically, the three types of final clarifiers used in the operation of small package wastewater treatment plants are described as hoppered, rectangular, or circular. Of the three basic configurations, the hoppered type is the most commonly used for package wastewater treatment plants (WWTPs).

The final clarifier follows the aeration basin in the treatment scheme, and the basic function of final clarification is to separate the mixed liquor suspended solids (MLSS) from the wastestream. The MLSS is formed in the aeration basin and contains millions of microorganisms and organic solids that can exert an excessive oxygen demand on the receiving stream if left untreated.

Final clarification must work in conjunction with the aeration basin, as a well-operating aeration basin will produce MLSS formed by small particles, which clump together to form larger ones. This process of forming larger heavier particles from smaller lighter ones is called flocculation. Flocculation forms a solids particle that is slightly heavier than water and can be removed from the wastestream by settling.

Effective separation of MLSS from biologically treated wastewater can only occur through proper design and operation of the final clarifier. Clarifiers function to physically remove solids by allowing them to settle to the bottom or float to the surface. Settled solids are either returned to the aeration basin to be remixed with the basins contents, or wasted from the treatment system either to a sludge holding tank or directly to waste disposal. Floating solids and scum are typically skimmed from the clarifier surface and returned to the aeration basin to mix with the biological solids. Proper removal of settled and floating solids from the wastestream produces a clear, treated wastestream that flows to a disinfection facility and eventually to the receiving stream.

The activated sludge process is inherent to producing mixed liquor solids that are only slightly heavier than water, and their settling rate is slow. Hydraulic currents in the clarifier will interfere with the settling process. The clarifier should be designed to minimize agitation and produce a still (quiescent) environment to allow suspended solids from the aeration basin to slowly settle. As the mixed liquor enters the clarifiers, the flow is dispersed in a manner to slow the velocity. Typically, with a package WWTP, the inflow enters at one end of the clarifier and a baffle is used to disperse and slow the wastestream. Circular clarifiers generally have a center feedwell with a circular feedwell baffle that serves the same purpose.

The treated water exits the clarifier using an overflow weir. The overflow weir is located as far as possible from the inlet to reduce short-circulating and allow for maximum settling time. As the treated water flows over the weir, a slight upward current is produced that settling solids must overcome.

If the overflow weir is not level, water will flow over only a portion of its length, producing a higher-than-recommended discharge velocity. If this current is great enough, solids could escape the clarifier. It is very important to establish a level weir so that water will flow evenly over its entire length. To prevent floating solids from leaving the clarifier with the treated water, an effluent baffle is installed that traps the floating solids and allows the treated water to flow under the baffle and over the effluent weir.

The clarifier environment must be kept as still as possible. Attention must be paid to the operation of surface skimmers, return or waste sludge pumps, and froth spray pumps so that their operation will not agitate the clarifiers' contents and interfere with settling.

As the MLSS enters the clarifier and solids begin to settle, a distinct solids-to-liquid interface develops. The height of settled sludge from the bottom of the clarifier is referred to as the *sludge blanket depth*. As the solids initially begin to settle, they start to form a blanket that acts as a filter to remove fine colloidal solids from the water as it slowly settles to the clarifier bottom. As the solids settle, the sludge blanket develops varying levels of solids concentration with the extreme bottom being the more concentrated.

Settled solids must be removed from the clarifier or the sludge blanket will continue to rise until it escapes over the effluent weir. The settled solids would also become anaerobic (without air) and denitrification would occur to release the solids in clumps to the clarifier surface. Maintaining a proper sludge blanket level by removing settled solids is very important to the effective operation of a final clarifier. The function of removing settled solids from the clarifier and returning them to the aeration basin is very important to the biological operation of the aeration basin. The returned settled solids help to maintain the microorganism population in the aeration basin.

The clarifier skimmer is designed to remove floating solids and scum from the clarifier surface. The skimmed water is typically returned to the aeration basin to mix with the MLSS.

Any nonbiodegradable floating particles such as plastic and wood should be manually removed from the clarifier.

Clarifiers, regardless of their type (hoppered, rectangular, or circular) have basically the same components. Each clarifier has an inlet structure designed to distribute the influent, at a reduced velocity, evenly into the clarifier. Each type of clarifier will have an effluent overflow weir and effluent baffle to control flow rates and prevent washout of floating solids. A sludge collector/removal system is provided to gather and remove settled solids from the bottom of the clarifier. While clarifiers may have different shaped and mechanical devices, they all exist to serve the same purpose: separation of solids from the treated water.

*T*YPES OF CLARIFIERS

Basically, there are three types of clarifiers used with package WWTPs. Their classification is based on their structural configuration: hoppered bottom, rectangular, and circular.

HOPPERED. Hoppered clarifiers are the most frequently used with package WWTPs. Hoppered clarifiers are standard in design with a baffled inlet structure, overflow effluent weir, effluent baffle, surface skimmer, and sludge removal pump. Typically, the skimmer and sludge removal use an air lift pump that draws air from the aeration basin blower system. The air pump has no moving parts other than the blower. The only maintenance of the air lift system is to ensure an adequate supply of air to each pump.

The main component of a hoppered clarifier that distinguishes it from other types of clarifiers is, as the name implies, the hopper bottom which consists of a small bottom with sloped wall surfaces that allows settled sludge to collect. The sloped surface should present an angle of less than 60 deg to the horizontal. Any angle greater than 60 deg will allow solids to accumulate on the wall surface and not be effectively removed. Slopes greater than 60 deg require manual scrapping of the clarifier walls. Figure 6.1 illustrates a typical hoppered clarifier.

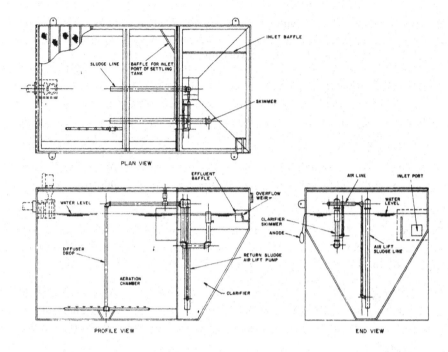

Figure 6.1   Typical hoppered clarifier.

The hoppered clarifier design can facilitate the accumulation of gravel, sand, or mud in the hopper bottom that can prevent proper operation of the sludge air lift pump. The hopper contents must be periodically monitored and cleaned of such debris.

RECTANGULAR. Rectangular clarifiers have the same inherent components as all clarifiers. The distinguishing feature of rectangular clarifiers is the manner in which sludge is collected and removed. Rectangular clarifiers typically use chain and flights carried on submerged sprockets, shafts, and bearing to gather sludge and transport it to a hopper located near the inlet end of the clarifier. The chain and flights are driven by a motor through a gear reducer. As the flights travel along the clarifier bottom moving settled sludge to the hopper, they also travel in the opposite direction on the clarifier surface, gathering floating matter and pushing it to a skimmer. Figure 6.2 illustrates a typical rectangular clarifier.

It is important that each of the flights in a rectangular clarifier be the proper length. If flights are too short, solids will build up on the outer edges and turn septic, rise to the surface, and be discharged to the effluent. If a flight is too long, it may rub the tank wall and break the flight; a broken flight could wedge in a position that results in breaking the remaining flights. Properly sized, each flight should have a clearance of approximately 25 to 50 mm (1 to 2 in.) between the end of the flight and the tank wall.

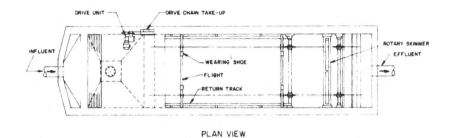

PLAN VIEW

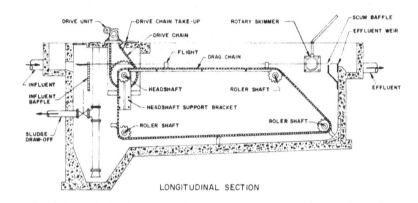

LONGITUDINAL SECTION

Figure 6.2   Typical rectangular clarifier.

Rectangular clarifiers do not experience the operation problems of hoppered clarifiers. Their flights are designed to continually scrape the bottom of the clarifier. This mechanical cleaning eliminates the ratholing problem of sucking a clear cone or channel in the sludge, frequently experienced with the hoppered clarifier. Clear water is than returned to the aeration chamber rather than the settled sludge. Figure 6.3 illustrates the problem of channeling sludge in a hoppered clarifier. Because the rectangular clarifier has vertical walls, the tendency of solids to cling to them is not as great as the sloped walls of the hoppered clarifier. The rectangular clarifier does have some disadvantages, however, when compared to the hoppered clarifier. The claims, flights, and drive motors must be maintained properly. By contrast, the air lift pump used commonly with the hoppered clarifier does not have any moving parts; therefore, it requires little mechanical maintenance. Rectangular clarifiers are used more often on plants whose design capacity is greater than 4.381 L/s (100 000 gpd).

CIRCULAR. Circular clarifiers are considered the most efficient of all types of final clarifiers but are typically used at larger WWTPs. Circular clarifiers

---

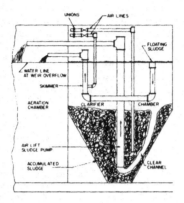

Figure 6.3   Channeling or ratholing of sludge in a hoppered clarifier.

provide a more even distribution of influent flow to the clarifier and typically provide more efficient use of the entire clarifier area for a distinct quiet settling zone.

Circular clarifiers use a motor-driven rotating rake mechanism to gather sludge and push settled sludge to a central collection hopper. Collected sludge is removed from the hopper using differential head (water elevation) to be either returned to the aeration basin or wasted to sludge-handling facilities. Floating material is removed by a surface skimmer that is attached to the sludge collection mechanism. The skimmer gathers floating debris and discharges it to a scum box for removal. Figure 6.4 shows a typical circular clarifier.

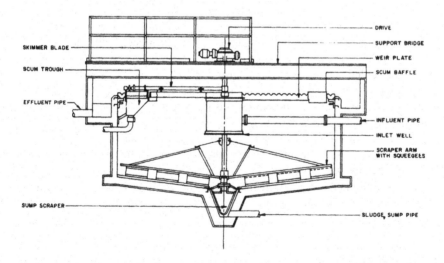

Figure 6.4   Typical circular clarifier.

DESIGN CRITERIA

Clarifiers are designed based on the following parameters.

- Surface settling rate,
- Weir overflow rate,
- Hydraulic detention time,
- Solids loading rate, and
- Inlet and outlet structures.

The listed design criteria are extremely important, as many operating problems can be attributed to a poorly designed clarifier.

SURFACE SETTLING RATE. A direct method of evaluating clarifier size is to use the surface settling rate, which compares the surface area of the clarifiers to the influent flow. The surface area remains constant, while the flowrate is variable. The following formula is used to calculate the surface settling rate:

$$\frac{\text{Surface settling rate}}{\text{m}^3/\text{m}^2 \cdot \text{d (gpd/sq ft)}} = \frac{\text{Influent flow (m}^3/\text{d [gpd])}}{\text{Clarifier area (m}^2 \text{ [sq ft])}} \tag{1}$$

Operators of package WWTPs can arrive at their actual surface settling rates by using the given formula. Design of the clarifiers should use the peak hourly flowrate to the clarifier. This concept is conservative in belief, but the purpose is to prevent solids from escaping the clarifier even during the most extreme hydraulic circumstance.

The recommended surface overflow rates for a package WWTP clarifier should be less than 12 222 L/m²d (300 gpd/sq ft) based on average flow, and less than 40 740 L/m²d (1000 gpd/sq ft) based on peak hourly flow. Surface settling rates must be addressed during design. The only operator control is to regulate influent flow to the clarifier. This, at times, can also be beyond operator control.

One common method to reduce peak flows is to construct a flow equalization basin ahead of the treatment plant. As discussed in Chapter 3, an equalization basin can provide a more uniform flowrate to the WWTP. Reduction in peak flowrates reduces the potential of solids being washout of the clarifier.

WEIR OVERFLOW RATE. Wastewater leaves the clarifier over weirs into effluent troughs. The length of the weirs in relation to the flow is important to prevent high velocities near the overflow weirs that may cause solids to be lost with the clarifier effluent. The weir overflow rate is the number of liters

---

per second (gallons per day) that flow over each meter (foot) of weir. The following is used to calculate the weir overflow rate:

$$\frac{\text{Weir overflow rate}}{\text{L/s [gpd/ft]}} = \frac{\text{Influent flow (L/s [gpd])}}{\text{Length of weir (m [ft])}} \qquad (2)$$

Based on average daily flows, the recommended weir overflow rate for package WWTP clarifiers is less than 0.4311 L/m·s (3000 gpd/ft), and less than 1.437 L/m·s (10 000 gpd/ft) for peak hourly flowrates Similar to the surface settling rate, weir overflow rates are hydraulic in substance and should be addressed during design. A WWTP operator can regulate influent flows using an equalization basin or increase weir length by adding additional effluent weirs. At times, this can be accomplished with a double-sided weir box as compared to a single weir.

Finally, the overflow weir must be level to operate properly; it was designed to have uniform flow over its entire length. If the overflow weir is not level, excessive discharge velocities, which will hinder settling, may occur.

Another common problem with overflow weirs is that they leak between the adjustable weir and the effluent trough. All water leaving the clarifier should flow over the discharge weir.

HYDRAULIC DETENTION TIME. Hydraulic detention time is the time required for wastewater to flow through the clarifier. The following formula is used to calculate the hydraulic detention time:

$$\frac{\text{Hydraulic detention time}}{\text{Hours}} = \frac{\text{Clarifier volume (gal)}}{\text{Influent flow (gpd)}} \times 24 \, (\text{h/d}) \qquad (3)$$

Detention time must be sufficient to allow for almost complete removal of the settleable solids; however, long detention times do not materially improve removal and may be actually harmful by allowing the sludge to become septic. In small package extended aeration treatment plants, the clarifier will have a detention time of four to six hours for average daily flow and two hours for peak flows. When calculating detention time for hoppered clarifiers, only the upper one-third (by height) may be used; the lower two-thirds are reserved for sludge storage.

SOLIDS LOADING RATE. Solids loading rate is the relationship between the mass of solids entering the clarifier and the surface area of the clarifier. The concentration of suspended solids often determines settling velocities; in activated sludge facilities this is an important parameter because the suspended solids loading to the clarifier is generally high. The settling rate, or speed of downward movement of the solids, slows down as the concentration of solids increases. If the concentration of suspended solids entering the clarifier becomes too high, the rate of downward movement of the particles caused by sludge drawoff from the bottom could be less than the rate at which solids enter and build up in the clarifier. As a result, solids could build up in the clarifier and be eventually discharged with the effluent. In a properly operating clarifier, solids will be removed from the clarifier at approximately the same rate they enter. The height of the sludge blanket should remain at approximately the same level. The following formula is used to calculate the solids loading rate.

$$\frac{\text{Solids loading rate}}{\text{kg/d/m}^2 \text{ (lb/d/sq ft)}} = \frac{\text{Suspended solids entering the clarifier, kg (lb)}}{\text{Clarifier surface area, m}^2 \text{(sq ft)}} \quad (4)$$

Calculation of the solids loading rate has become a very important consideration in clarifier design. The mass of suspended solids entering the clarifier may be calculated by determining the MLSS, return sludge flowrate, and influent flowrate to the plant. The MLSS is expressed in milligrams per liter; the return sludge and influent flowrates are expressed in million gallons per day. Suspended solids entering the clarifier each day are equal to

$$\frac{\text{Solids Loading Rate,}}{\text{kg/m}^2 \cdot \text{d}} = \frac{\text{MLSS, mg/L} \times (\text{Influent flow, m}^3/\text{d} + \text{Return sludge flow, m}^3/\text{d}) \times \left(0.001 \frac{\text{kg/m}^3}{\text{mg/L}}\right)}{\text{Clarifier surface area, m}^2} \quad (5)$$

$$\frac{\text{Solids Loading Rate,}}{\text{lb/d/ft}^2} = \frac{\text{MLSS, mg/L} \times (\text{Influent flow, mgd} + \text{Return sludge flow, mgd}) \times \left(8.34 \frac{\text{lb/mil. gal}}{\text{mg/L}}\right)}{\text{Clarifier surface area, ft}^2} \quad (6)$$

The solids loading rate should not exceed 97.64 kg/m²·d (20 lb/d/sq ft) on a daily average flow or 170.87 kg/m²·d (35 lb/d/sq ft) on a peak hourly flow. Excessive solids loading to the clarifier can lead to higher-than-recommended sludge blanket levels and increase the possibility of solids escaping the final clarifier.

Solids loading to the clarifier can be somewhat controlled by the plant operator by adjusting MLSS concentrations. Mixed liquor suspended solid concentrations are discussed in more detail in Chapter 5.

INLET AND OUTLET STRUCTURES. Inlet and outlet structures, including appropriate baffles, are important to clarifier efficiency. Inlet baffles reduce the velocity of the mixed liquor entering the clarifier and distribute the flow horizontally and vertically. Inlet baffles should prevent short-circuiting and turbulence through the clarifier. The clarifier should remain quiet when the aeration chamber is being mixed; if disturbances appear in the clarifier while

---

the aeration chamber is being mixed, the inlet baffle is not working properly. Modification of the baffle should be considered.

Discharge baffles are designed to prevent floating solids from leaving the clarifier. The discharge baffle will typically rise approximately 152.4 mm (6 in.) above the liquid level and extend approximately the same distance below. A common manufacturing error is to use the same design for the discharge battle as the inlet baffle. The inlet baffle should extend at least 152.4 mm (6 in.) below the bottom of the port from the aeration chamber to the clarifier to reduce the liquid velocity. However, if the effluent baffle extends that deep below the liquid surface, it may be in the sludge blanket, causing solids to escape from the clarifier even though the supernatant may be clear. Thus, the effluent baffle should extend only 152.4 mm (6 in.) below the liquid level.

The recommended distance between the inlet and outlet should be a minimum of 3.048 m (10 ft), unless special measures are taken to prevent short-circuiting.

CLARIFIER OPERATION

For a package WWTP to achieve optimum treatment efficiency, the final clarification unit must effectively separate the biological solids from the mixed liquor. If these solids are not separated properly and removed from the clarifier in a timely manner, operating problems will result, causing an increased load on the receiving stream and a decline in treatment efficiency. The most important function of the final clarifier is to maintain the wastewater quality produced by the preceding processes.

Because the purpose of the clarifier is to separate floating and settleable solids from the liquid, it should be operated properly. The operator should inspect the clarifier and its operation thoroughly during every visit. Ideally, the plant should be visited daily; however, because of the unavailability of an operation, lack of money, or location, many small package extended aeration plants are visited as infrequently as twice a week. The quality of the effluent and regulatory requirements will determine the frequency of plant visitation.

RETURN SLUDGE RATES. The return sludge rate will directly affect clarifier operation. The operator must adjust the rate of sludge withdrawal to maintain a proper inventory of solids in the clarifier. Sludge blankets are typically recommended at a maximum of one-third of the side water depth of the clarifier and a minimum of 152.4 mm (6 in.) from the clarifier bottom. Maintaining sludge blanket levels greater than this increases the possibility of solids washout. Maintaining sludge blankets less than 152.4 mm (6 in.) is inefficient use of the sludge withdrawal system, as the concentration of solids will be low and more water than solids will be removed from the clarifier.

An excessively high return sludge rate will increase velocities at the inlet to the clarifier and in the clarifier itself. This will disrupt the sludge blanket and

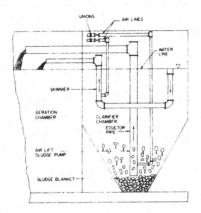

Figure 6.5   Properly operating hoppered clarifier.

cause solids to be swept over the effluent weir. An insufficient return rate
may allow solids to remain in the clarifier too long, causing them to become
septic, float to the surface, and possibly flow out with the effluent. Gas bubbles
in the clarifier are another indication that sludge is remaining in the clarifier
too long. Floating sludge and gas bubbles could also indicate equipment
failure. If an underwater chain or flight is broken on a mechanical clarifier,
sludge can not be removed properly from the clarifier.

Figure 6.5 illustrates a properly operating hoppered clarifier. The sludge
returns to the aeration chamber at approximately the same rate is enters the
clarifier. A minimum sludge blanket is maintained in the clarifier; however,
the return rate is not excessive. If the return rate is too high, currents could
cause disruptions in the clarifier.

A problem in hoppered clarifiers caused by high return sludge rates is
illustrated in Figure 6.3. A high return rate may create a rathole, cone, or
channel in the sludge blanket; a return sludge that is relatively clear is a good
indication of channeling. Channeling may be corrected by squeegeeing the
walls of the clarifier and reducing the return sludge rate. The operator should
squeegee a hoppered-type clarifier at each visit to the treatment plant.

Sludge return rates are typically recommended to range from 50 to 150%
of influent flow. The majority of small package WWTPs do not have the
facilities to monitor the influent and return sludge flowrates. Maintaining
effective sludge return rates for package WWTPs involves monitoring the
sludge blanket levels and monitoring the concentration of the returned sludge.
Sludge blanket levels should range from a minimum of 152.4 mm (6 in.) to a
maximum of one-third of the side water depth. The return sludge concentration
should be considerably thicker than the MLSS concentration typically in the
2 to 4 times greater range.

SKIMMING AND SCUM REMOVAL. Floating solids are removed from
the clarifier by the skimmer. Maintenance of the skimmer consists of keeping

the equipment clean, adjusted, and in proper operation. Skimmers on hoppered clarifiers are somewhat ineffective in removing floating solids, especially solids that collect in corners. Improperly adjusted, the air lift skimmer can create enough turbulence to hinder settling, and on heavily loaded plants can draw the sludge blanket to the surface. On hoppered clarifiers, the top of the air lift skimmer should be approximately 6.35 mm (0.25 in.) below the liquid surface when the blowers are operation. The air lift skimmer should use the minimum amount of air needed to pump the liquid back to the clarifier. The skimmer air valve should be set at the required minimum to provide the maximum available air to the diffusers and return sludge pump. If the air valve is wide open, the skimmer will rob most of the air from the diffusers and return sludge pump.

Skimmers on rectangular and circular clarifiers are mechanically operated and are typically attached to the settled sludge removal equipment.

Skimming equipment should be closely monitored to where it is only used when floating solids are to be removed. Constant operation of a skimmer can return excessive volumes of water to the aeration basin that, in turn, reenters the clarifier as influent flow. This excessive flow increases the hydraulic loading to the clarifier.

FROTH SPRAYS. Some package extended aeration plants have a submersible pump located in the clarifier to pump clear liquid to the froth spray in the aeration chamber. A hose bibb is installed on some froth spray system so that the clear liquid may be used to wash down the plant or to mix chemicals. If at all possible, one should have a potable line, protected by a proper backflow preventer, installed near the plant and used for washdown and mixing chemicals. **The submersible froth spray pump should not be operated unless it is absolutely necessary.**

WATER TEMPERATURE. The temperature of the mixed liquor that enters the clarifier influences the settling rate of the floc. As the temperature decreases, water becomes denser, increasing the resistance to floc settling. At a water temperature of 26.67°C (80°F), the settling rate of the flow will be almost 50% faster than at 10°C (50°F). Water is at maximum denseness at 4°C (39.2°F). The lower liquid temperature during the winter months will typically result in lower removal efficiencies and poorer effluent quality.

TROUBLESHOOTING

As with any biological or mechanical systems, problems can occur. The following provides a discussion of commonly encountered problems and a listing of remedies.

FLOATING SOLIDS. Floating solids are commonly referred to as clumping, ashing, or rising sludge. Floating solids typically occur because of a high sludge age (too many solids in the treatment system or too long of a solid retention time in the final clarifier). The following are remedies for floating solids.

(1) Decrease solids inventory by increasing sludge wasting rates,

(2) Increase sludge return rates to remove settled sludge quicker, and

(3) Check for dead spots or areas of accumulated sludge where sludge is not being collected for removal.

SOLIDS OVERFLOW. Excessive solids losses in the final clarifier can be the result of hydraulic overload or can occur because of the type and characteristics of the biological solids present. The following are remedies for solids overflow.

(1) Use settleability test to determine sludge age. For young sludge, increase solids inventory by decreasing the sludge wasting rate to produce an older sludge which tends to settle faster. For old sludge, increase wasting rates to decrease solids inventory.

(2) Check for short-circuiting.

(3) Calculate detention time and check for hydraulic overloading.

RISING SLUDGE. Rising sludge is quickly recognized as rising clumps of feathery sludge that are easily shattered at the clarifier surface. Rising sludge is caused by denitrification in which nitrates in the wastewater are reduced to nitrogen gas. Denitrification occurs when the sludge become anaerobic. As nitrogen gas accumulates, the sludge mass becomes buoyant and floats to the surface. To remedy the problem of rising sludge, increase sludge removal by increasing sludge removal rates.

FOULING OF EFFLUENT WEIRS. An accumulation of solids or algae on the weir surface can cause short-circuiting with the clarifier, creating excessive velocity currents that pull solids over the effluent weir. The following are remedies for fouling of effluent weirs.

(1) Thoroughly clean the weir surface, and

(2) Use chlorine solution to prevent algae buildup.

PLUGGING OF SLUDGE WITHDRAWAL LINES. Plugging of return sludge lines is typically caused by too high of a solids concentration in the return sludge or a buildup of mud and sand in the clarifier bottom. The following are remedies for plugging of sludge withdrawal lines.

(1) Withdraw sludge faster and more frequently, and

(2) Manually remove mud and sand.

The clarifier is an extremely important part of the wastewater treatment plant. If it does not operate properly, the clarifier will cause a poor quality effluent. The clarifier must remain quiet so that settleable and floating solids can be removed from the mixed liquor. The skimmer and return sludge pump must be operated properly to remove solids from the clarifier in a timely manner.

Chapter 7
Sludge Wasting

*O*VERVIEW

A package extended aeration wastewater treatment plant that receives the proper amount of food, has the correct pH, is supplied with plenty of air, and has good mixing will not produce a clear effluent indefinitely. The microorganisms in the treatment plant must be periodically removed or wasted from the activated sludge process to maintain a high-quality effluent. Microorganism growth can be dramatic; a single fecal coliform bacteria commonly found in a treatment plant can multiply to produce a biological mass equal to the weight of the Earth in 24 hours if enough food were present and the microorganism was not swimming in its own waste!

The extended aeration process is designed so that the microorganisms are operating in a declining growth mode where this astronomical growth capability is offset with limited food and endogenous metabolism (the microorganisms consume their internal food and energy supplies to stay alive) and microorganism death. For the extended aeration process, the sludge remains in the plant for 30 to 40 days at design loading and longer if the plant is underloaded. This means that 1/30 to 1/40 of the activated sludge inventory must be wasted daily to prevent the plant from becoming choked with solids. If sludge is not wasted routinely, solids may accumulate in the plant and be lost over the clarifier discharge weir, resulting in poor effluent quality. Therefore, it is essential to have an effective solids management program as part of the plant's process control program. An effective solids management program for an extended aeration plant should

(1) Establish optimum mixed liquor suspended solids (MLSS) concentrations and grams (pounds) of total solids inventory in the aeration tank and clarifier;

(2) Provide information to the operator on MLSS or solids inventory in the aeration tank and clarifier; and

(3) Use a generally accepted process control strategy to maintain optimum solids inventory levels, such as constant MLSS in aeration tank, mean cell residence time (MCRT), or food-to-microorganism ratio (F/M). The solids levels in the digester or holding tank are not included in the activated sludge solids management program because these have been removed from the activated sludge process. These solids levels must also be monitored to demonstrate compliance with regulatory requirements and ensure cost-effective sludge disposal; this is addressed at the end of this discussion.

SLUDGE WASTING

There are three generally accepted activated sludge process control strategies, which can be used to establish the optimum sludge wasting rate. These are

(1) Maintaining a constant MLSS concentration, 30-minute settleability solids level, or centrifuge spin test level in the aeration tank. Typical optimum ranges for extended aeration are MLSS between 2000 and 6000 mg/L, 30-minute settleability solids between 10 and 30%, and centrifuge spin solids between 2 and 20%.

(2) Maintaining a constant MCRT, which is defined as the grams (pounds) of solids under aeration divided by the grams (pounds) of solids wasted via sludge wasting or that are lost in the effluent. Typical optimum MCRT range is 30 to 40 days for extended aeration. It is important to include effluent solids in the MCRT calculation for extended aeration because this can be a significant contribution to MCRT for lightly loaded plants.

(3) Maintaining a constant F/M, where F/M is defined as grams (pounds) of influent biochemical oxygen demand (BOD) per day divided by grams (pounds) mixed liquor volatile suspended solids (MLVSS) in the aeration tank. Typical optimum F/M range is 22.68 to 68.04 g (0.05 to 0.15 lb) BOD per gram (pound) MLVSS per day.

Each process control strategy has advantages and disadvantages; for further discussion on these aspects, numerous wastewater training manuals and technical books on biological wastewater treatment are available. For a listing of these, contact the state regulatory agency or local operator organization. The key to successfully operating a package treatment plant is to maintain a process control strategy; do not change the wasting rate more than 10% per day and be patient. Research has shown that it may take as long as three sludge ages to return to steady-state operation after a process change, i.e., if the plant is operating with a 30-day MCRT, it may take 90 days to return to steady-state.

From an experience standpoint, when recovering from a process upset, process performance typically gets worse before it improves; therefore, the operator must be patient when dealing with process changes.

Package plants are typically operated by maintaining a constant 30-minute settleability solids level in the aeration tank. When a settleability or centrifuge spin test is used for solids management, it is worthwhile to occasionally double-check against aeration MLSS test as performed by a laboratory. The operator should check more often if process performance is erratic. The settleability solids test can be misleading if sludge quality changes, and there may be a higher true solids inventory than indicated by the settleability solids test when sludge settling characteristics change.

From a process control standpoint, sludge should be wasted continuously to maintain steady-state conditions; for larger plants, sludge wasting can be automated to accomplish this. In smaller plants, continuous sludge wasting is typically not feasible because of manual steps required to divert waste sludge flow and related constraints. Sludge wasting in package plants is typically performed one to three times per week. The following general procedure is useful for wasting sludge in a package plant.

(1) Always ensure adequate space is available in the aerobic digester or sludge holding tank before wasting sludge. This can be accomplished by turning the air off in the aerobic digester or holding tank for at least one hour to allow sludge to settle. Clear liquid can be supernated or pumped back into the plant to make room for the new volume of sludge to be wasted. Some operators combine supernating with wasting and let the new waste sludge push the clear liquid back into the plant. This method is not preferred because it typically stirs the sludge and digester solids already wasted may be returned to plant, resulting in deterioration of effluent quality. If the aerobic digester or sludge holding tank has no means to pump clear liquid back to plant, the previously mentioned method is the only option.

(2) Run a 30-minute settleability solids test on the aeration tank mixed liquor to determine the aeration tank solids inventory. Remember that only 10% of that reading is theoretically to be wasted per day. It is recommended to perform the 30-minute settleability solids test three to five times per week to monitor solids inventory and sludge wasting needs.

(3) Most package plants waste sludge from the clarifier return activated sludge (RAS) line, so turn off the RAS air valve or pump for approximately 30 minutes before wasting sludge to build a thickened sludge to waste to aerobic digester. This can be done while waiting for the aerobic digester to settle in step 1.

(4) Once sufficient room is available in the aerobic digester or sludge holding tank, waste sludge for approximately 15 minutes by opening the waste sludge airlift valve or splitter box, or by turning the waste sludge pump on to divert the RAS flow to the aerobic digester or sludge holding tank. This is an arbitrary time and the proper wasting time needs

to be established for each plant by determining the wasting rate necessary to maintain the optimum 30-minute settleability solids levels in the aeration tank.

(5) After sludge is wasted, it is imperative to return valve and pump settings to their normal position. The RAS flow must be returned to the normal setting to maintain the activated sludge process and the air turned back on in the aerobic digester or sludge holding tank to prevent the digested sludge from going septic and producing foul odors. During the sludge wasting cycle, the RAS sludge concentration is generally thicker when compared to normal operation, so be careful to give the RAS pump sufficient time to get back to its normal flowrate before leaving the plant.

(6) Eventually, the aerobic digester or sludge holding tank solids will become too thick to supernate and there is no more room in the tank for the waste sludge volume. When this occurs, sludge solids should be removed from the plant by draining tank contents to an on-site sludge dewatering system or hauled off by tanker truck for proper disposal. Only during extended periods, when the plant is lightly loaded, can the sludge be returned to the aeration tank (i.e., at the end of the school year or camping season), providing complete oxidation of all organic material. After the sludge has been totally oxidized, the plant is often turned off until the next operating season. If the digester or sludge holding tank is pumped down, it should be refilled with sufficient water to minimize any uplift pressures from groundwater in the surrounding soil.

Sludge disposal

The thickened sludge in the aerobic digester or sludge holding tank must be properly disposed of in accordance with state and federal regulations. Sludge drying beds are not favored in most locations, and the typical method of sludge removal is by on-site dewatering facilities or liquid hauling via tanker trucks.

Properly stabilized aerobic digester sludge can be beneficially reused as part of an environmentally sound conservation program. Stabilized sludge is typically referred to as biosolids to distinguish this valuable material from hazardous sludge.

The U.S. Environmental Protection Agency (U.S. EPA) 40 CFR 257, 403, and 503 regulations went into effect in 1992 and address proper handling, testing, and recordkeeping for biosolids disposal and reuse. Additionally, numerous states have adopted biosolids regulations that may be more stringent than U.S. EPA regulations, so it is best to contact the state regulatory agency for guidance in compliance with these regulations. Generally, biosolids that are beneficially reused must not exceed (1) ceiling concentration criteria for nine metal concentrations, (2) fecal coliform levels, and (3) stability level (vector attraction reduction potential as measured by specific oxygen uptake rate or other parameter or method). Biosolids from a municipal treatment plant generally are capable of meeting the nine metal concentrations levels, unless there are significant industrial discharges. However, biosolids typically require further processing by lime stabilization or other methods to meet regulatory criteria for fecal coliform and vector attraction reduction.

The biosolids regulations, although appearing complex, have been developed to open the door for beneficial reuse of this recyclable material, while still protecting the public and the environment. State and U.S. EPA regulatory agencies and numerous professional biosolids management companies can serve as valuable resources ensuring the proper disposal of biosolids.

Chapter 8
Disinfection

GENERAL

Disinfection is the process of killing disease-causing microorganism (pathogens) that remain in effluent. Effluent can come from the final clarifier or additional treatment units such as sand filtering or polishing ponds. A disinfection unit may be the last treatment unit in the wastewater flow before discharge to a water-body, unless dechlorination is provided. In small package extended aeration treatment plants, disinfection is typically accomplished by either chlorination or UV radiation.

All organisms are not destroyed during the process. This differentiates disinfection from sterilization, which is the destruction of all organisms. In wastewater treatment, the three categories of human enteric organisms of greatest consequence in producing disease are bacteria, viruses, and amoebic cysts. Diseases caused by waterborne bacteria include typhoid, cholera, paratyphoid, and bacillary dysentery. Diseases caused by waterborne viruses include poliomyelitis and infectious hepatitis.

CHLORINATION

The most common method of disinfection used in package extended aeration plants is chlorination. To destroy pathogenic bacteria chlorination, there must

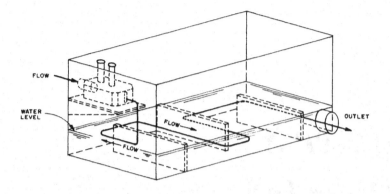

Figure 8.1 Typical chlorine contact tank.

be sufficient contact time between bacteria and chlorine. A chlorine contact tank provides the required time (Figure 8.1). Contact time is provided by sizing the tank with sufficient volume for the required contact period and by using baffles in the tank to eliminate short-circuiting.

The chlorine tank is typically attached to the clarifier if the plants are constructed of steel. Concrete chlorine contact tanks are typically separate units. Concrete tanks are virtually maintenance-free. Steel tanks must be painted periodically with coal tar epoxy paint or other recommended corrosion protection that is applied according to the manufacturer's instructions to prevent deterioration caused by the highly corrosive chlorine.

Occasionally, solids will carry over into the chlorine contact tank from the clarifier. These solids must be periodically removed by taking the tank out of service. A two-compartment tank is typically needed unless chlorination is not needed year-round.

Regulators agencies typically require 15 minutes chlorine contact time during peak hourly flow or approximately 900 L of chlorine tank volume for each liter per second (10.4 gal for each 1000 gpd of design flow). In addition to contact time, regulatory agencies also limit the amount of chlorine that can be discharged in effluent. This concentration in effluent varies among regulatory agencies, but is typically in the range of 0.2 to 0.5 mg/L.

Chlorinators

Three different types of chlorination systems are commonly used in small wastewater treatment plants: gas, liquid, and solid tablet or pellet chlorinators. Each has its advantages and disadvantages. Regardless of the type used, the operator should verify that the chlorinator is sized for the peak hourly design flow of the treatment plant. Sources of replacement parts should be identified and parts should be readily available. If replacement parts are not available, another make of chlorinator should be considered. Finally, the designer should consider the type of chlorine supply available when selecting chlorinators.

GAS. Gas chlorinators use either a small portion of effluent from the treatment plant or potable water as the chlorine carrier. This liquid is then pumped through an injection nozzle or diffuser, where the chlorine gas and liquid are mixed. The liquid containing chlorine is then applied to the treatment plant effluent. Because this type of chlorinator uses pure chlorine, which is toxic, the manufacturer's operating instructions must be adhered to rigidly. **Under no circumstances should one attempt to operate a gas chlorinator without first reading the manufacturer's operating manual.** Figure 8.2 shows a typical gas chlorinator.

While a gas chlorinator may cost more initially than other chlorinators, there can be savings in the long run. The principal advantage of using the gaseous form is the infrequence of handling the chlorine supply. With 70-kg (150-lb) cylinder service, a 0.5 L/s (100 000 gpd) plant would probably have to be replaced once a month. Chlorine in the gaseous form is the least expensive and storage does not cause a loss of strength. If a gas chlorinator is used, the proper safety devices (ventilation, self-contained pressurized gas masks, and leak detectors) should be provided. In addition, the system should be inspected by the appropriate regulatory agency.

Figure 8.2 shows the functioning of a gas chlorinator schematically. For safety purposes, chlorine is not under pressure. A vacuum (created by water being forced under pressure through the ejector nozzle) pulls a resilient

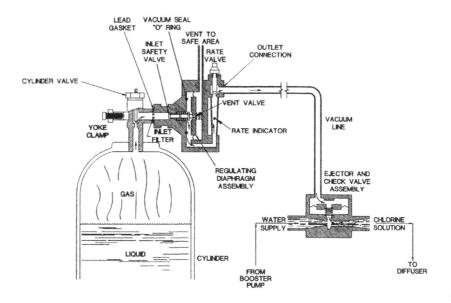

Figure 8.2   Schematic of cylinder-mounted, vacuum-operated gas chlorinator.

diaphragm, which pushes open a springloaded inlet/safety-shutoff valve. The vacuum draws gas from the cylinder, through the chlorinator, then through high-strength vacuum tubing, and into the ejector. There, it mixes with effluent water that is discharging through the ejector, and is carried to the diffuser, which passes it into the wastewater being treated.

Should anything happen to cause a break in any part of the system, gas does not leak out. Air leaks in, and the vacuum is lost. With no vacuum to pull it open, the spring onto the inlet/safety-shutoff valve snaps the valve shut, stopping the gas supply immediately and automatically.

An adjustable feed-rate valve and indicator are built into the system to allow the flow of gas to be manually adjusted and observed. Figure 8.3 shows a complete gas chlorinator installation.

LIQUID. Three common types of liquid chlorinators are used on small treatment and use positive displacement pumps because they deliver a definite amount of solution with each stroke. The pumps are calibrated in gallons of liquid pumped in 24 hours. Figure 8.4 shows a typical liquid chlorinator.

Sodium hypochlorite (liquid laundry bleach) and calcium hypochlorite (powder high test hypochlorite [H.T.H.]) are used as the chlorine sources for liquid chlorine pumps. Liquid sodium hypochlorite is recommended for use because it is readily available in local supermarkets and is easy to use. Its chief disadvantage is that it contains only 5.25% available chlorine and is, therefore, an expensive source of chlorine. Commercial-grade sodium hypochlorite is available at 15% strength. Because the shelf life of liquid bleach is limited, it is important to use a supplier with fresh stock.

Calcium hypochlorite is a white powder that contains 65 to 70% available chlorine. When dissolved in water for use in liquid chlorinators, a powdery

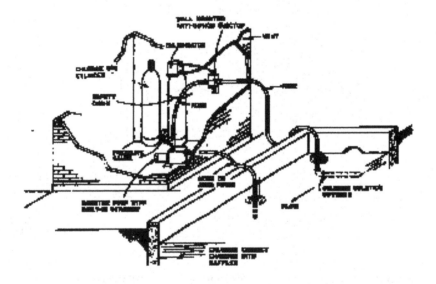

Figure 8.3 Gas chlorinator installation.

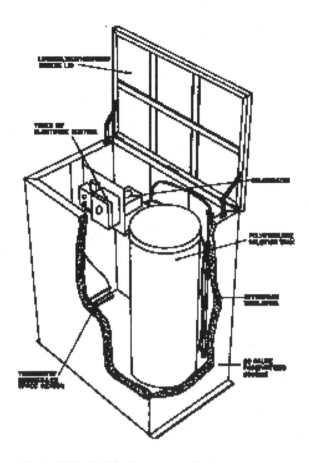

Figure 8.4   Typical liquid chlorinator installation.

residue of calcium carbonate and calcium hydroxide will remain undis-
solved. The residue formed must be kept from the chlorinator pump to avoid
plugging chlorine feed lines and fittings. This is done by dissolving as much
of the calcium hypochlorite as possible in a separate mixing container and
letting it set quietly over night. The following day, the clear liquid is poured
into the chlorine solution tank. The residue remaining in the mixing container should be discarded at a landfill; it is not to be dumped into the
wastewater treatment plant. The cost of available chlorine in calcium
hypochlorite is less than that of sodium hypochlorite, but is more labor-
intensive to apply.

SOLID TABLET OR PELLET. Chlorinators using tablets or pellets of
calcium hypochlorite have become popular for use in small wastewater
systems. The tablets are place in vertical stacks that are slotted to allow
wastewater to flow through the automatic feed as they dissolve in the effluent
stream from the plant. The amount of chlorine fed is controlled by the number

---

of stacks containing tablets (in multiple stack chlorinators), and by a weir that controls the water level in the chlorinator.

One major advantage of the system is that chlorination occurs only when there is flow. A major disadvantage is that the tablets are relatively expensive and they can jam inside the tubes. The same brand of tablet as the manufacturer of the chlorinator is typically required, and the residual chlorine concentration in the effluent is difficult to control during periods of high or low flows through the treatment plant. Care must be taken when filling the stacks. If tablets are dropped into the tube, the bottom of the tube could be knocked off or the tablets may crack. To load the tube, the operator must lean it on its side and slowly slide tablets down the tube. No more than six tablets should be placed in a stack at one time. If more are used, the lower tablets will leave a residue that will not allow the upper tablets to fall into the wastewater stream. Figure 8.5 shows a typical tablet chlorinator unit and servicing schedule.

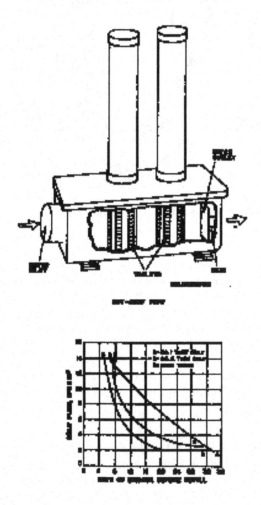

Figure 8.5 Typical tablet chlorinator unit and servicing schedule.

CONTROL SYSTEMS

Whether the plant is equipped with a gas or liquid chlorinator, the proper amount of chlorine must be added to the effluent. The operator can control the amount of time the chlorinator operates and the amount of chlorine injected. Wastewater flows fluctuate during the day. Flows in the morning and evening will be several times greater than the flow received during other times of the day. More chlorine is needed during the high (peak) flows than during low-flow periods. A typical flow diagram is shown in Figure 8.6.

The treatment plant will experience the hourly flow variation shown in Figure 8.6 only if the flow is gravity fed. In many installations, the flow must be pumped or lifted to the plant. Some designers have used two pumps, each capable of pumping the peak hourly inflow. In the case shown in Figure 8.6, the 24-hour flow is approximately 4.86 L/s (111 000 gpd [77 gpm]) and the peak flow is approximately 10.95 L/s (250 000 gpd [174 gpm]). Therefore, that plant would receive flow at the rate of 10.95 L/s (250 000 gpd), but intermittently. In recent years, with the advent of variable frequency drives, each pump would be capable of following the inflow.

The above considerations make it difficult to select a chlorine control system. Automatic control for small plants may not be cost effective. Therefore, for a gravity inflow system, the dosing rate must be set manually. In some cases, it may be possible to manually reset the dosing rate two to three times per day. The dosings would be determined through trial and error to obtain the

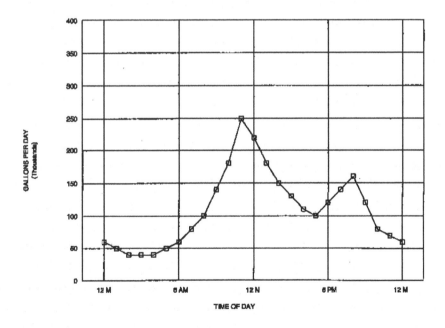

Figure 8.6   Typical flow diagram (1 gpd = 264.2 m³/d).

---

required residual. For those systems that pump all the influent at a constant rate, the chlorinator could be tied into the on/off switch of the pump.

Another control common to both gas and liquid chlorinators is to vary the amount of chlorine injected when the chlorinator is operating. For gas chlorinators, this is done by adjusting the chlorine feed rate on the controller. For liquid chlorine feeders, the rate can be adjusted by the control knob on the pump. Liquid chlorinators have one other adjustment that can be made: the strength or concentration of the chlorine solution.

For a new system or for initial startup of a new chlorinator, the proper settlings are typically found by trial and error. The manufacturer's operating instructions will provide an initial recommendation, but the settings will have to be fine-tuned. The flow pattern in Figure 8.6 is typical for residential flow, but will differ from school or campground flows.

For liquid chlorinators, the chlorine solution strength may also be varied. Following the manufacturer's recommendation, enough chlorine solution should be prepared to fill the chlorine feed tank. The pump settings are set and the residual is checked. If the residual is too high, the chlorine feed stock should be diluted. If the residual is too low, more chlorine should be added to the stock. If the manufacturer's instructions are not available, one can try 8.0 L (1 gal) of bleach (5.25% available chlorine) in 84 L (10 gal) of water, 1:10 by volume, or 0.5 kg (1 lb) of H.T.H. dry powder (70% available chlorine) in 84 L (10 gal) of water. **Remember: the dry powder must be premixed and only the clear liquid is to be added to the solution tank.**

Liquid chlorinators must be cleaned periodically. This can typically be done by pumping white vinegar through the unit to dissolve any mineral deposits. Chlorine solution tanks must also be cleaned. Even if potable water is used to mix the chlorine, a residue will form from the minerals in the potable water. The chlorine solution tank should be cleaned periodically as experience indicates.

DECHLORINATION

In cases where chlorine residuals may have potential toxic effects on aquatic organisms, dechlorination of treated effluent is practiced. Dechlorination may be accomplished by reacting with a reducing agent such as sulfur dioxide (SO_2).

Sulfur dioxide is available commercially as a liquefied gas under pressure in steel containers with capacities of 45, 68, and 907 kg (100, 150, and 2000 lb). Sulfur dioxide is handled in equipment in a similar way as in standard chlorine systems. When added to water, SO_2 reacts to form sulfurous acid (H_2SO_3), a strong reducing agent. The H_2SO_3 dissociates to form hydrogen sulfite (HSO_3), which will react with free and combined chlorine, resulting in formation of chloride and sulfate ions. The reaction with the total chlorine residual is accomplished in less than two minutes.

The principal elements of a SO_2 system are the SO_2 containers and feeders (sulfonators), solution injectors, diffuser, mixing chamber, and interconnecting piping. It is similar to a chlorine gas system.

ELECTROLYTIC GENERATION OF HYPOCHLORITE

Hypochlorite can be generated on site electrolytically from salt (NaCl). Figure 8.7 is a view of the hypochlorite generator and a schematic. The major components are the cell water softening, brine tank and pump, dilution water controls, a microprocessor control package, and the hypostorage tank.

The hypochlorite strength is 1%. It takes approximately 1.16 kg (2.55 lb) salt, 51.63 L (13.64 gal) soft water, and 2.5 to 2.75 kW hours of electricity to make 0.45 kg (1 lb) equivalent of chlorine.

Salt is dissolved in softened water and forms a saturated brine solution. The brine is diluted with additional softened water and then flows into the cell. As it passes through the cell, it is exposed to a direct current (DC), which converts the salt into sodium hypochlorite solution and hydrogen gas.

Hypogenerating systems are designed to maximize reliability and safety. When the microprocessor detects that the hypochlorite level in the storage tank is low, it first conducts system-wide safety checks before starting the electrolytic cell. The flows of brine and dilution water are allowed to rise to normal operating rates, while the interlock system verifies that the hydrogen dilution system is operational.

When the control system is satisfied that all of the components are operating within specified parameters, the rectifier is started and electric current is applied to the cell. Production of the 1% hypochlorite solution begins. The hypochlorite solution is transferred to the storage tank, where the hydrogen gas separates from the solution and is diluted well below lower explosive limits by air, which is blown into the tank by the air dilution system. The hydrogen/air mixture, which has been diluted to safe levels, is then force-exhausted outside the building. The 1% sodium hypochlorite solution in the storage tank can be safely dosed using a wide variety of application equipment or stored for future use without the concern for degradation associated with commercially produced hypochlorite solutions.

ULTRAVIOLET DISINFECTION

Ultraviolet disinfection has been used for the disinfection of water since the early 1990s and applied more recently for the disinfection of wastewater. Ultraviolet systems (Figure 8.8, Tipton Environmental International, Batavia, Ohio) incorporate a UV lamp(s) submerged in the process stream. The lamp(s) emit UV energy, which is passed through the process liquid as it flows through the system. Organisms within the process liquid are bombarded with the UV energy and are killed by its effect.

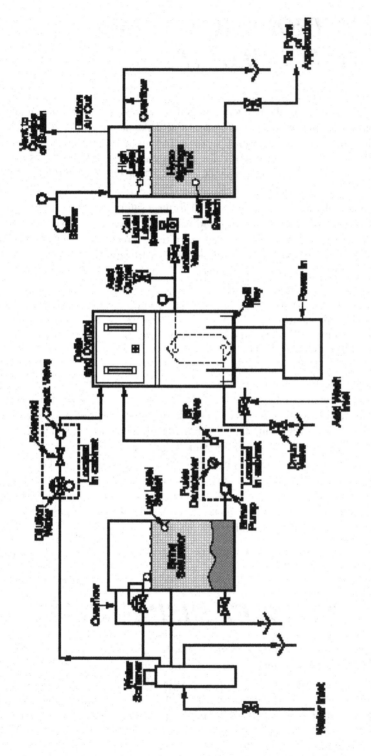

Figure 8.7   Typical hypochlorite generating system.

Figure 8.8   Typical UV disinfection unit (Courtesy Trojan Engineering).

Ultraviolet energy is naturally occurring and is produced by the sun in great quantities. The earth's atmosphere absorbs most of the sun's UV energy, protecting the earth's surface from overexposure. Ultraviolet energy is emitted in that portion of the electromagnetic spectrum known as the optical range, which encompasses UV, visible, and infrared wavelengths. Microbiological organisms are composed of a single cell, 10% of which consists of deoxyribonucleic acid (DNA). The DNA defines the operations within a cell, including reproduction. Absorption of the UV energy by the cell causes photochemical damage to the reproductive process of the DNA within the cell and stops cell reproduction. A cell that cannot reproduce is considered deactivated.

Maintenance of the UV unit consists of cleaning the quartz or Teflon tubes and replacing bulbs. Some units are equipped with either automatic or manual wipers that clean tubes without removal of the UV lamps. Small units use about the same amount of electricity as a light bulb. Care must be taken to ensure that power is disconnected when bulbs are cleaned or changed. Exposure to UV radiation is harmful, and looking at the lighted bulbs for long periods of time will damage the eyes.

Chapter 9
Operating Tests and Routine Maintenance

INTRODUCTION

Most large wastewater treatment plants are operated by highly trained individuals who use sophisticated equipment and testing procedures. Smaller plants are typically operated by personnel who have limited training and use elementary testing procedures and simple equipment. Nonetheless, small plants can be operated successfully.

Parameters that can be monitored easily and that give a good picture of the treatment plant's condition are settleability, pH, color, dissolved oxygen, and residual chlorine (if chlorine is used for disinfection). If portable test kits are used, everything will fit into an approximately 11.36-L (3-gal) plastic bucket. Required test equipment includes the following:

- An approximately 11.36-L (3-gal) plastic sample bucket with rope,
- Dissolved oxygen test kit,

- pH test kit,
- Two calibrated quart jars,
- A residual chlorine test kit, and
- Elbow-length rubber gloves.

One additional piece of equipment needed to operate the plant successfully is a squeegee that is long enough to reach the bottom of the plant.

SETTLEABILITY

An indicator of good settling is a test that measures how well the biological floc will settle. There are two such tests: settleability and settleometer.

SETTLEABILITY TEST. The settleability test, or 30-minute settling test, should be considered the major process control test for small package plants. A 1000-mL graduated cylinder or a calibrated mason jar may be used to determine the percent of settled sludge by volume. Figure 9.1 shows a simulated settleability test using a quart jar. The jar has been calibrated by markings on a piece of adhesive tape at 13-mm (0.5-in.) increments from the bottom to where the jar begins to curve. The marks are then labeled from 0 to 100%, at 10% intervals.

A sample from the aeration chamber (taken after blower has operated for 10 minutes) is allowed to settle in the jar for 30 minutes, after which the percent of settled solids is determined. The samples should be taken at the same location in the aeration chamber and at the same time of day to allow for comparison of tests taken on different days. The sample should not be taken near the plant influent or near a return sludge line. When the settleability test is conducted, the quart jar should be placed in the shade, on a level surface, and away from any vibration caused by the blowers.

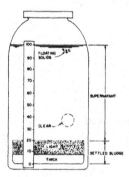

Figure 9.1 Simulated 30-minute settleability test using modified quart canning jar.

The operator should observe the settleability test for the first five minutes. How the sludge settles is just as important as the final amount that settles. During the first five minutes of the settleability test, a healthy sludge should compact slowly, forming a screening blanket and squeezing clear liquid from the sludge. A good settling sludge will settle at 20 to 50% of its original volume after 30 minutes. A problem may exist if the sludge settles quickly, leaving finer particles in the supernatant, even though the final percent solids reading is within the acceptable range. A rapidly settling sludge, cloudy supernatant, and dark brown color typically indicates an old sludge with a large amount of inorganic solids; in this case, increasing sludge wasting would than be recommended. If the settleability test results are less than 20% and the supernatant is cloudy, but the sludge settles slowly and the color is light brown, than a young sludge is probably present. In this situation, the system could simply be lightly loaded.

At times, the settleability test results will be above the recommended range. This condition could be caused by either too much or too little sludge in the system. The particular problem can be determined by the 50% dilution test. If the wastewater treatment plant is just experiencing startup, microorganisms are growing rapidly and have not developed enough weight to settle well. The sludge will have a light brown color and very little setting will occur after 30 minutes. If this is the case, then sludge wasting should be reduced or eliminated until the microorganism population produces a good settling floc. At that time, sludge wasting could be initiated as described in Chapter 7.

A high sludge reading could also be produce by an old sludge. If sludge wasting is inadequate, the sludge will become old, denser, and will compact easily. Initially, the percent solids may seem to decrease. If inadequate wasting is continued, the old sludge will eventually accumulate, even though it compacts well. The percent solids reading will continue to increase above the recommended range.

DILUTION TEST. Because the plant effluent will be less than optimum, it is important for the operator to determine which condition exists so that corrective measures may be taken. The 50% dilution test will provide this information; this test is conducted by filling one-half of the quart jar with a sample from the aeration chamber, and filling the remaining one-half with unchlorinated clarifier effluent. Clear tap water should not be used for dilution, as it tends to make the sludge rise. The typical settleability test is the run on this sample. If the 50% diluted sample does not settle any better than in the first test, then the sludge is young. If, however, the diluted sample settles significantly better than the original test, the sludge is old and sludge wasting should be initiated. It should not be assumed that the original sample settled poorly simply because there were too many solids in the system.

SETTLEOMETER TEST. Another control test procedure similar to the settleability test is the settleometer test. This test is conducted with a sample from the aeration chamber. The sample is placed in a 1000-mL beaker or a quart jar and allowed to settle for 60 minutes, with readings taken every 5 minutes for the first 30 minutes and every 10 minutes during the second 30 minutes. Settling results from the settleometer will indicate what is taking place in the clarifier.

The settleometer test readings are then plotted on graph paper, with the time variable on the horizontal axis and the settled sludge reading on the vertical axis. The slope and shape of the curve indicates the sludge quality. Figure 9.2 shows an ideal sludge settling curve for a plant that is operating properly. The ideal settling curve does not guarantee a clear discharge. The settleometer test simply indicates how the sludge should settle in the clarifier. If the clarifier is designed properly and all units are operating effectively, the effluent from the clarifier should be clearer than the supernatant in the settleometer. If the effluent is not clearer, there is a problem in the clarifier.

After an ideal settling curve has been developed, the operator must try to maintain it; however, there is no cause for alarm if the plant occasionally deviates from the ideal curve. The operator has little control over what enters the plant. If possible, the public should be informed of what should and should not enter the system. Assuming that the plant is not receiving a toxic material, it is important that the operator recognizes possible reasons the settling curve is above the ideal curve and possible reasons why it is below. Figure 9.3 shows a condition where the actual settling curve is above the ideal curve. As was discussed in the Settleability Test section, the high settling curve is caused by either too little or too much sludge. The 50% dilution test is used to determine which condition applies; the procedures described previously should then be followed.

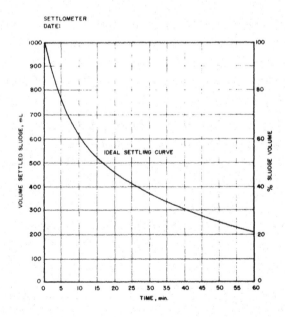

Figure 9.2   Ideal settling curve.

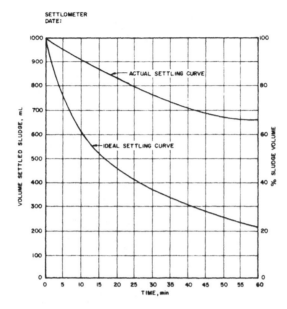

Figure 9.3   High sludge settling curve.

At times, the actual settling curve may be less than the ideal, as illustrated in Figure 9.4. This condition could result for several reasons. In the figure, the dashed settling curve indicates an old sludge that contains a large amount of inorganic solids, settles very rapidly, and has a dark brown color. This condition can typically be corrected by increasing sludge wasting.

The other actual settling curve is also below the recommended range. If the sludge does not settle rapidly and has a light brown color, the plant may simply be lightly loaded. Sludge wasting, in this case, would not be required. How the sludge settles and what color it is will determine whether or not the plant is lightly loaded. Flow data can also help determine if the plant is lightly loaded.

The condition of the settled sludge after 60 minutes is significant. All sludge that begins to rise after 60 minutes may be overoxidized. A properly oxidized sludge will not rise to the surface until two to four hours after the test begins. If the sludge begins to rise at the 60-minute mark, sludge wasting is required.

Although many parameters can be used to control the operation of a package extended aeration plant (i.e., food-to-microorganism ratio [F/M], depth of sludge blanket, dissolved oxygen, return sludge rate, and sludge volume index), the settleability and settleometer tests provide the required information, yet do not require a laboratory or a laboratory technician. It is important, however, that the small plant operator understand how to interpret test results.

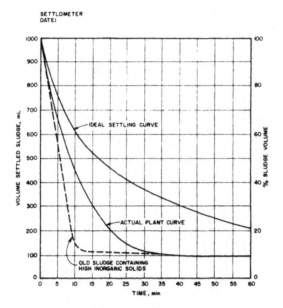

Figure 9.4   Low sludge settling curve.

pH

As noted before, the pH in the aeration chamber should be between 6.5 and 8.5 for microorganisms to grow. A pH meter or a small liquid color comparator test kit may be used to test the pH. Although the color comparator is excellent in determining pH for operational control, its results may not be accepted by the appropriate regulatory agency. The regulatory agency should be contacted to determine whether color comparator results are acceptable. If the results are unacceptable, the effluent pH must be determined by a pH meter. The pH meter takes time to calibrate and must be maintained according to manufacturer's recommendations, whereas the color comparator is simple to use but does have some limitations.

High suspended solids or high chlorine residuals in the testing samples will interfere with the pH test. Samples taken from the aeration chamber must be allowed to settle for approximately 10 minutes, at which time the pH of the supernatant can be checked. Effluent samples with chlorine residuals greater than 1.0 mg/L must be dechlorinated. If a color comparator is used, its range should be at least one pH unit above 8.5 and one pH unit below 6.5. The wide range pH test kit, with a range from 4.0 to 10.0, is popular among small plant operators. Whether a pH meter or color comparator kit is used, the instructions furnished with the test equipment must be followed carefully. Test equipment should be clean at all times and chemical reagents should not become contaminated.

Once the pH value has been determined, the operator must enter that value on the operating log (Table 9.1). Any time a test is conducted, it should be entered on the operation log. Poor record keeping is one of the biggest deficiencies for small package extended aeration plant operators. A history of the plant operation is often the key to solving current operating problems.

COLOR

The color of the aeration chamber is one of the quickest ways to check the system operation; the color should be brown (similar to coffee with cream). If the aeration chamber is this color and has a musty odor similar to a damp basement or mushrooms, the sludge is probably healthy. If the color is gray, the plant is not receiving enough air. Possible reasons for this include the following: the plant is receiving too much food, control time clocks are not allowing the aerators to operate enough, diffusers may be partially plugged, or the plant may have received some toxic material. A black color accompanied by a rotten egg odor indicates that the plant is septic. When this occurs, the plant should be placed on constant aeration until the light brown coffee color returns. A septic plant is typically the result of poor attention by the operator. Although aerator and power failures, plugged diffusers, and toxic material will cause the plant to go septic, the most common cause is neglect.

Other colors that may be observed occasionally are white, red, and purple. A white aeration chamber occurs when the plant is extremely lightly loaded. The sludge, in this case, is completely oxidized and only ash remains. Feeding the plant is not recommended. A red color may be encountered when the plant is overaerated, and a filamentous bacteria called *leptothrix* is present. The sludge will settle poorly and thick matty foam will form. A purple color is not natural, but has been encountered when iron removal water systems, which use potassium permangante to regenerate the media, use the wastewater system to dispose of backwash and flushing water. Therefore, the operator should know what is connected to the wastewater system.
Remember: the package extended aeration treatment plant is designed to treat *domestic* waste only; anything else can cause an operating problem.

DISSOLVED OXYGEN

Dissolved oxygen is one of the most important ingredients of the mixture in the aeration chamber. Oxygen is necessary for aerobic bacteria to use organic material; without it, septic conditions will result. The amount (concentration) of oxygen that can be dissolved in the mixed liquid is temperature dependent; the colder the water, the greater the amount of oxygen that can be dissolved (Table 9.2). The oxygen concentration is measured in milligrams per liter

Table 9.1   Typical operating log.

SEWAGE TREATMENT LOG						PROJECT				CAPACITY & TYPE			LOCATION			

Date	Aeration Chamber									Settling Tank				Chlorine Solution						Effluent		
				30 min mixed liquor solids (%)	Blower Setting (min)					Depth of Clear Water (mm inches)	Sludge Return (color)	Surface Scum		Strength		Pump Setting		Timer Setting or No. of Samuril Tubes		Solids (%)	Res. Cl₂ (mg/L)	
(See reverse side for remarks)	Time	Water Meter (gal)	DO (mg/L)	pH		On	Off	Color	Odor					Liters (gal) Bleach	Liters (gal) Water	Scale:						TESTS BY

ALL TESTS Conducted Twice Weekly

Table 9.2   Solubility of oxygen in fresh water versus temperature.

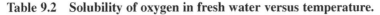

°C	°F	O₂ (mg/L)	°C	°F	O₂ (mg/L)
0	32.0	14.6	26	78.8	8.2
1	33.8	14.1	27	80.6	8.1
2	35.6	13.8	28	82.4	7.9
3	37.4	13.5	29	84.2	7.9
4	39.2	13.1	30	86.0	7.5
5	41.0	12.8	31	87.8	7.5
6	42.8	12.5	32	89.6	7.4
7	44.6	12.2	33	91.4	7.3
8	40.4	11.9	34	93.2	7.2
9	48.2	11.6	35	95.0	7.1
10	50.0	11.3	36	96.8	7.0
11	51.8	11.1	37	98.6	6.9
12	53.6	10.8	38	100.4	6.8
13	55.4	10.6	39	102.2	6.7
14	57.1	10.4	40	104.0	6.6
15	59.0	10.2	41	105.8	6.5
16	60.8	10.0	42	107.6	6.4
17	61.6	9.7	43	109.4	6.3
18	64.4	9.5	44	111.2	6.2
19	66.2	9.4	45	113.0	6.1
20	68.0	9.2	46	114.8	6.0
21	69.8	9.0	47	116.6	5.9
22	71.6	8.8	48	118.4	5.8
23	73.4	8.7	49	120.2	5.7
24	75.2	8.5	50	122.0	5.6
25	77.0	8.4	51	123.8	5.5

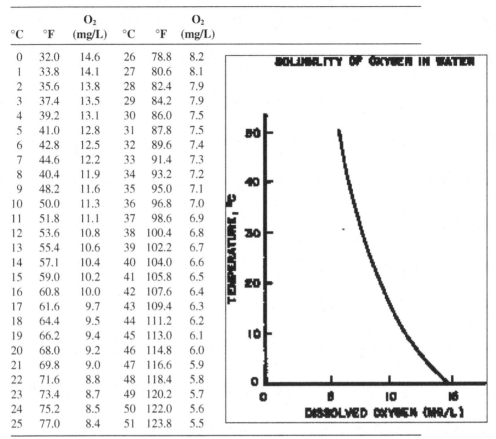

(mg/L); the ideal oxygen concentration in the aeration chamber is 2 mg/L. The dissolved oxygen in a clarifier should be at least 1 mg/L to prevent septic conditions, which would result in gas bubble formation, rising sludge, and unpleasant odors. When wastewater flows are erratic, aeration will be difficult to yield the ideal oxygen concentration. It is, therefore, better to have a dissolved oxygen concentration that is greater than 2 mg/L than to let it become completely depleted, as fewer unpleasant consequences result.

DISSOLVED OXYGEN METER. The test for dissolved oxygen is performed on the mixed liquor from the aeration chamber for operational control and may have to be conducted on the plant effluent for regulatory control. Because oxygen is required in the aeration chamber at all times, plants that are aerated intermittently must be sampled near the end of the "off" aeration period, to indicate the minimum oxygen dissolved in the water. The best method for testing dissolved oxygen in the aeration chamber is to use a dissolved oxygen meter. If the probe is placed at mid-depth in the aeration chamber, the dissolved oxygen concentration can be read directly. The meter eliminates the need to obtain a sample from the aeration chamber, a common source of error. Sampling is ineffective when

- Samples are inadvertently agitated, which induces oxygen;
- The sample is taken near the surface, which does not give a true indication of the oxygen content in the aeration chamber; or
- The sample is not tested immediately.

If a dissolved oxygen meter is unavailable, the proper sampling equipment to obtain a sample at mid-depth of the operation chamber should be used. The American Society for Testing and Materials (ASTM) Special Technical Publication No. 148-1 and the U.S. Geological Survey (USGS) Water Supply Paper No. 1454 (1960) are good references for determining proper equipment.

DROP TITRATION COLORIMETRIC TEST. Satisfactory dissolved oxygen data may also be obtained from a drop titration colorimetric test kit. The operation should follow the test kit directions, test equipment must be kept clean, and all test data must be recorded on an operating log.

DISSOLVED OXYGEN AMPOULES. Although not approved for regulatory test data, dissolved oxygen ampoules containing premeasured reagents may be used to test dissolved oxygen in the aeration chamber. These ampoules are quick, inexpensive, and easy to use. Although this method is not suited to give research-type results, it has proven successful for operational purposes. Color and odor may also be used to determine that dissolved oxygen is sufficient. If the color of the mixed liquor is like coffee with cream and there are no disagreeable odors, the dissolved oxygen is probably satisfactory.

Because operation of the aerators costs money, the operator should aerate no more than required. However, small systems do not have the personnel or operating controls that large plants have; therefore, the operator should be cautious when fine-tuning the system to the ideal dissolved oxygen concentration. Dissolved oxygen concentrations of 5 mg/L in the aeration chambers of small plants are common. Unlike large plants, where each milligram per liter of dissolved oxygen may cost hundreds of dollars in electrical cost, the operating cost for each milligram per liter of dissolved oxygen in small package extended aeration plants is relatively low.

FLOW

Flow in small plants is typically expressed in liters per second (gallons per day). Because flow information is important for plant operation, a flow-measuring device is required. For raw wastewater, an open-channel type flowmeter, such as a parshall flume, is desirable so that wastewater solids will be passed without clogging. At the plant effluent, a V-notch weir is satisfactory. It is also highly desirable to have a flow recorder that displays instantaneous flowrates and cumulative flow. The chart will show peak and low-flow periods that may radically affect not only the clarifier, but overall plant operation. Flows that exceed the hydraulic capacity of the clarifier or other process units can be detected. A single peak-flow period occurring daily can cause solids washout in the clarifier and prevent the activated sludge process from becoming established. If this condition exists, peak flows must be reduced by increasing plant size, adding flow equalization, or diverting flow elsewhere. Daily flow records should be maintained, with readings taken at the same time each day. Although peak flows will not be reflected, water meter readings will give a good indication of the average flow to the plant.

RESIDUAL CHLORINE

If possible, the residual chlorine test should be run daily to determine whether sufficient chlorine is being fed to consistently disinfect the plant effluent. The required chlorine residual range is typically between 0.2 and 1.5 mg/L; other values may be required if bacteriological tests show disinfection to be adequate or inadequate. The effectiveness of disinfection at a given residual level will vary, depending on the concentration of solids or nitrogen. Changes in the concentrations can cause the chlorine residual to change with a given feed rate. It is important, therefore, to monitor residual chlorine often and make feed adjustments as needed. The regulator agency should be contacted to determine what procedures may be used to detect chlorine. The most common field test kit used is the *n*-diethyl-*p*-phenylenediamine (DPD) kit; this kit is sold commercially by several companies. As always, the instructions on the kit should be followed carefully, the kit should be clean, and all test data should be recorded. Samples for the chlorine residual test are to be taken at the point nearest to the receiving stream or at the exit point of the chorine contact tank.

Some regulatory agencies may require that the plant effluent contain little or no chlorine residual. In that case, the operator must apply a dechlorinating agent to the plant discharge. If dechlorination is required, the operator should consider UV radiation rather than chlorine to disinfect the discharge. As noted in Chapter 8, the UV method causes no side effects to the environment, as nothing is added to the plant effluent. A variety of styles and sizes of UV units are available. The method may require that the effluent be filtered before disinfection.

TYPICAL OPERATOR VISIT

Ideally, the operator should visit the plant daily. Some of the major items that should be checked include the communitor, bar screen, equalization tank, aeration chamber, clarifier, chlorine contact tank, and mechanical equipment.

COMMINUTOR. Check for sticks, rags, and rocks that may have become caught in the comminutor. **Remember: Turn off power before cleaning.** Perform maintenance as required by the owner's instructions. **Do not use hands to remove debris.**

BAR SCREEN. Remove material from bars and place on drying rack. Place dried material in plastic bag and dispose at landfill.

EQUALIZATION TANK. If system has an alkalization tank, remove any floating material and dispose with bar screen debris. Check diffusers to ensure they operate properly. Inspect floats and observe operation of the unit.

AERATION CHAMBER. Perform settleability and setteometer test; note color of aeration chamber; note rate of return sludge, color, and odor; check pH and dissolved oxygen; inspect froth spray nozzles and rolling action produced by diffusers (remove and clean every three months); record all test data; and wash down as needed.

CLARIFIER. Gently squeegee all sides; adjust return sludge if needed; adjust skimmer if needed; remove nonbiodegradable floating material (i.e., rags, sticks, plastic, grease balls, watermelon seeds); check packing nuts on all air valves; inspect overflow weir (must be level); inspect clarifier for high currents; check depth of clarity (approximately 305 to 457 mm [12 to 18 in.] is common) with a Secchi disk; and finally, record all results.

CHLORINE CONTACT TANK. Check chlorine residual and make feed adjustments if needed; check chlorine supply for chlorinator; and inspect chlorine contact tank for solids.

MECHANICAL EQUIPMENT. Check time clocks for correct time of day and day of week. Ensure that air is supplied to the plant daily. Establish a routine lubrication schedule for each piece of mechanical equipment, as recommended by the manufacturer. Record the date of each lubrication in both a maintenance log and on a tag placed on the piece of equipment. Finally, check air filters and clean or replace as needed; check air relief valves; and inspect air piping for leaks.

*R*EFERENCES

ASTM (unknown) Manual on Industrial Water and Waste Water, Special Technical Publication No. 148-1; American Society for Testing and Materials: Philadelphia, Pennsylvania (out of print).
USGS (1960) Methods for Collection and Analysis of Water Samples, Water Supply Paper No. 1454; U.S. Geological Survey: Reston, Virginia.

Chapter 10
Advanced Unit Processes

INTERMITTENT SAND FILTERS

Intermittent sand filtration of wastewater originated at the turn of this century as a method of raw wastewater treatment. Because of large land area requirements and high costs of construction, operation, and maintenance, its use diminished until the 1970s, with the advent of more stringent removal requirements for small wastewater treatment works, particularly package plants. In smaller applications, intermittent surface sand filters are cost-competitive with other polishing processes, because they are simple to understand, operate, and maintain, and they offer a large surface area to handle secondary plant effluent surges. In addition, intermittent sand filters can produce excellent effluent. A report entitled "Low Maintenance Mechanically Simple Wastewater Treatment Systems" (Rich, 1980) noted that intermittent sand filters produced effluent with average suspended solids and biochemical oxygen demand (BOD) concentrations of 10 mg/L and ammonia-nitrogen levels of 1 mg/L. Figure 10.1 shows an intermittent surface sand filter.

The intermittent surface sand filter removes suspended solids and organic matter by physical straining and biological oxidation. Therefore, the disinfection unit is more efficient when located after the intermittent surface sand filters.

Intermittent surface sand filters are typically dosed by mechanical means. Dosing is required by many regulatory agencies because filters may operate more efficiently if they are flooded with 75 to 100 mm (3 to 4 in.) of waste-water in 10 to 15 minutes, and then allowed to rest for several hours. For many smaller systems with sporadic flow, gravity-fed filters produce an excellent effluent. One filter is used until it starts to plug, and then the secondary effluent is diverted to the other filter, allowing the partially plugged filter to dewater. Once the plugged filter has dewatered and the solids mat has dried, the mat must be removed. Typically, the mat and approximately 13 mm (0.5 in.) of sand are removed with a flat shovel or similar device. The run time of any filter is a function of the influent solids; therefore, the better the package plant effluent, the longer a filter can be used and the lower the maintenance cost.

When one-half of the original sand depth is lost as a result of plugging, sand should be added to restore the level to its original depth. Only the mat

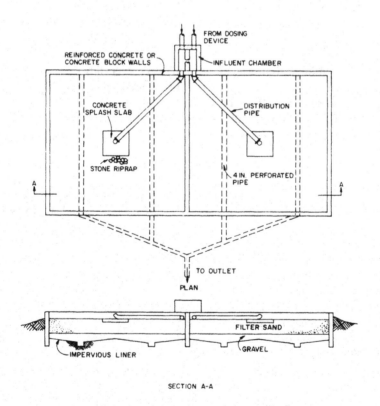

Figure 10.1   Intermittent surface sand filter.

and dirty sand should be removed before new, clean sand is added. When the sand is delivered or inspected at the supplier, a clean glass jar should be filled to the halfway point with the proposed sand, and water should be added. Next, the sand is stirred vigorously and the liquid clarity can be observed; if it is muddy, the sand should be rejected. Use of dirty sand will cause high suspended solids in the filter effluent, resulting in permit violations and potential early plugging of the filter. A type of sand used for sandblasting has proven to be among the best for initially clean sand as it contains few fines.

It is important to remove all vegetation from filters before root systems become established. Filters constructed with clean sand typically do not have a significant weed problem.

The dosage frequency may need to be increased in very cold climates because of the possibility of freezing. By increasing the dosage frequency, the amount dosed will decrease and the wastewater will remain on the filter for less time.

Each intermittent surface sand filter should not exceed 60 m^2 (600 sq ft) because of problems associated with uniform dosing—specifically the need for large dosing tanks that may require aeration and large pumps. Depending on the prevailing requirements and size and type of media used, filtering rates of package plant effluent of 400 to 1000 L/m^2·d (10 to 25 gpd/sq ft) are used. The sand should be a minimum of 460-mm (18-in) deep, have an effective size of 0.3 to 0.9 mm (0.012 to 0.031 in.), and have a uniformity coefficient less than or equal to 4.0. Three layers of gravel, each 75-mm (3-in.) deep, support the sand; they are typically 3 to 6 mm (0.125 to 0.25 in.), 6 to 10 mm (0.25 to 75 in.), and 20 to 40 mm (0.75 to 1.5 in.), respectively. The gravel is laid (larger sizes to the bottom) on an impermeable base that is sloped to the underdrains.

Duplex dosing capability is recommended. When a common dosing line is used, check valves will be needed. Drilling a weep hole on the discharge side of the check valve should allow drainage and prevent freezing; otherwise, separate dosing lines should be considered. Low points between the pump and distribution box must also be avoided if freezing is a possibility.

Distribution pipes extend from the distribution box and require manual alternation when a filter is plugged. The pipe should discharge onto a concrete pad or a sheet of 3-mm (0.125-in.) thick aluminum plate, at least l-m (3-ft) square, with large gravel hand placed around the pad to prevent local sand scour. The filter walls must be watertight, stable, and allow a minimum of 0.3 m (12 in.) of freeboard. A bypass line is recommended for maintenance purposes or emergencies, if permitted by local authorities.

MICROSTRAINERS

Microstraining is a physical straining process used for the separation of solids from a liquid suspension by passing the liquid through a porous membrane, defined as *screening media*. Separation occurs through the retention of solids,

---

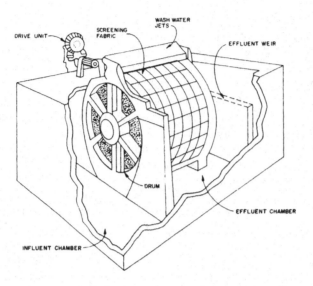

Figure 10.2   Typical microscreen unit.

either directly by capture on the screening media, or indirectly by capture on the "biomat" formed on the screening media from the previous capture of solids.

Typical microstrainer applications include wastewater, lagoon and secondary effluent polishing, prescreening in water treatment plants, combined sewer-storm water overflow treatment, primary treatment, industrial water conditioning, product recovery, and waste treatment applications.

A microstrainer consists of screening media engaged to the periphery of a rotating horizontal drum. Flow enters through the open inlet end of the drum and is dispersed radially through the screening media. Solids are retained on the screening media during the dispersion phase. Removal of solids from the screening media occurs during the backwashing phase. Spray water is supplied from backwash nozzles located on headers above the drum. Typically, the backwash header is constantly operational, thus producing a continuous regenerative screening media. Screened solids are washed into a backwash trough located inside the drum, and are discharged to the solids handling processes or are recycled to the head end of the biological processes. Figure 10.2 illustrates a typical microscreen unit.

RAPID SAND FILTERS

FUNDAMENTALS. Among the most common methods used in small wastewater systems for removal of BOD and suspended solids beyond levels achievable by secondary treatment is the rapid sand or mixed media filter. Design criteria from larger municipal and industrial filters have been success-

---

fully adapted to smaller system. Packaged filters that have fully automatic backwash capability are available from several manufacturers. Operation and maintenance instructions supplied by the manufacturer should be followed to achieve optimum performance. An efficient packaged automatic filter designed to follow an activated sludge plant should include the following process elements.

Constant Flow Through the Filter. Most filters are designed to operate at a particular flowrate. Flow equalization can be provided either ahead of the secondary treatment unit, or in a surge basin that is just ahead of the filter.

Sedimentation Ahead of the Filter. This configuration is particularly true if chemical coagulation is included ahead of the filter. However, this process is more desirable if it follows an activated sludge plant to prevent occasional sludge bulking, which quickly plugs the filters. One of the best devices is a module of either inclined parallel plates or tube settlers with a chamber below to store settled solids. Provisions should be made to return these solids to the secondary process or disposal.

Single or Multiple Filters. Two or more filters are desired to allow for uninterrupted operation while one filter is backwashing. Filter media may consist of a single grade of sand, or may include graded sands and possibly anthracite coal. The media support structure may consist of graded gravels, with the coarsest bottom layer surrounding a collection header and laterals. Some filters use a perforated plate that supports graded gravels and sand; others use a plate containing nozzles with small openings to support the sand.

Early filters for secondary plant effluent polishing used media designed to remove turbidity from potable water supplies. These media included sand of a very small diameter (0.5 mm or smaller). When applied to effluent polishing, where the goal was to achieve approximately 10 mg/L suspended solids, these media quickly plugged, resulting in short filter runs, unnecessarily high effluent quality, solids buildup in the system, and frequent filter bypassing. If a filter uses media in this size range and experiences such problems, it may be possible to replace the media with coarser material (i.e., 1.0 mm or larger). This method will result in longer filter runs, while still achieving the desired effluent quality. The procedure should not be attempted, however, without consulting the manufacturer or an engineer experienced in wastewater filtration.

In general, the multimedia bed consisting of graded sands and anthracite provide higher filtering capacity than single or media beds. Larger particles are trapped in the coarse first layers, leaving smaller particles to pass through the finer layers deeper in the bed. In this manner, more of the media is used for filtering, and longer runs are achieved before backwashing. In single media beds, only the top 25 to 50 mm (1 to 2 in.) are used effectively for filtering.

Backwashing Facilities. Adequate backwashing capability is required for rapid sand filters. Pumps and piping should be sized to deliver a minimum of 10 L/m²·s (15 gpm/sq ft) of filter area, with throttling capability for final flow

---

adjustment. Backwash rates up to 17 L/m^2·s (25 gpm/sq ft) may be required. Sufficient storage of treated water for backwash should be provided to accommodate a minimum of five minutes of backwash at full flow. Proper flowrate is important to ensure that the media is lifted and gently agitated, but not washed out. This action is necessary to loosen entrapped solids and allow flushing, while retaining the media. In multimedia beds, backwash rate is particularly critical because of the low density of the anthracite layer, which tends to be washed out easily.

An air scrubbing system is provided in some filters as part of the backwash cycle. Air is introduced to either the underdrain laterals or a separate distributor in the upper media layers; this air agitates and scrubs the media to loosen solids. Air is supplied by a positive displacement blower or compressor and is timed to occur at the beginning of the backwash cycle, but before the backwash pumps begin operating. This action provides violent agitation without washing out media.

Backwash is typically initiated, either manually or automatically, in response to increased head loss caused by plugging of the filter. Various mechanisms are used depending on the manufacturer's specifications.

Backwash-to-Waste Storage. Storage should be provided for backwashed water, or backwash-to-waste. Ideally, a separate chamber is used, with float switch-operated pumps metering the water back to the head of the secondary treatment unit during low-flow periods.

Clearwell. Tankage should be available to store filtered water. The tank should be sized to provide sufficient detention time for disinfection as specified by regulatory agencies, if it is to be used as a chlorine contact chamber. The clearwell should also be sized to accommodate a minimum of one full backwash cycle for each filter in the system. The residual chlorine levels required for disinfection will typically not cause harmful effects on secondary biological treatment units, if backwashed water is recycled to the head of the plant.

Bypass. A bypass should be provided from the top of the filter to the chlorine contact chamber to prevent upset of the secondary treatment system in the event that the sand filter plugs or freezes.

TYPICAL OPERATION. For illustration purposes only, the operation of a typical rapid sand filter is described here. This unit consists of a settling chamber, dual filter chamber, clearwell chamber, and backwash storage and surge chamber. Figure 10.3 illustrates the flow pattern through this packaged unit. Effluent from the secondary treatment unit flows into the settling chamber. The secondary effluent, which carries suspended solids, is then subjected to additional clarification by sedimentation. The liquid is then pumped to a filter feed weir trough, which allows flow to be distributed evenly over the length of each filter bed. The influent than falls approximately

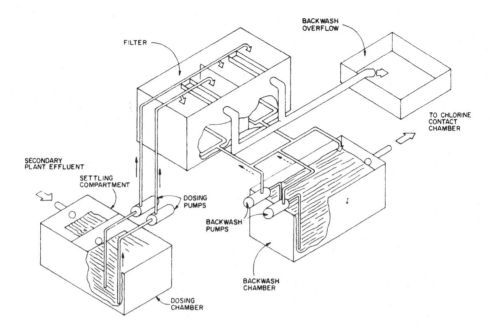

Figure 10.3 Typical rapid sand filter.

0.75 m (30 in.) to a minimum 150-mm (6-in.) pool of water maintained over the filter media in each bed.

This 150-mm (6-in.) pool of water over the filter media is maintained by the elevation of the discharge pipe from each filter bed. These pipes discharge into the clearwell feed chamber. Any increase in water level over the media forces the water through the media and into the backwash feed chamber. Solids are retained on the filter. The discharge lines contain automatic valves that close for testing and backwash of the filter equipment and beds. Filtered water collects in the backwash chamber until overflowing into a chlorine contact chamber where it is then discharged. Stored water in the backwash chamber is used to test backwash pumps and valves, in addition to providing a complete backwash cycle.

As solids are trapped in the filter, causing the water flow to be obstructed, head loss is increased. The water level over the media then rises to a predetermined height of approximately 0.3 to 1 m (12 to 35 in.) above the top of the media. At this point, the preprogrammed backwash cycle is initiated automatically, and performs as follows.

- Energizes relays that prevent both filters from backwashing simultaneously;
- Energizes the air scour blowers and opens automatic valves, allowing low-pressure air to flow into the filter bed (air-scrubbing occurs for approximately two minutes and ends before the backwash pumps are turned on);

- Energizes the backwash pump for approximately five minutes; and
- It returns the filter to normal operation.

A test cycle is provided by an adjustable time clock that operates all backwash equipment briefly at preset intervals of 0 to 24 hours. The operator may wish to override the automatic backwash initiated at a specific head loss, and use the timer to backwash periodically (once or twice daily). Water used for backwash then flows to the backwash surge chamber for storage. Time clock or float-operated pumps discharge at this point to the plant headworks during low-flow periods. The effluent is disinfected by adding chlorine to the clearwell or a separate chlorine contact tank.

Packaged filter unit controls are premounted and tested at the factory and contain control logic for all filter operations, in addition to contractors and motor starters. The manufacturer should furnish a complete operations and maintenance manual, including detailed drawings, control schematic, and operational descriptions. These instructions should accurately describe the proper operation of the facility, including installation and initial startup procedures.

ROUTINE OPERATION. The filter unit will enter the routine or normal operation stage when flow into the unit becomes stabilized. The unit operation should then become a matter of monitoring the effluent for any deterioration in quality, general cleaning or housekeeping, and inspecting and lubricating mechanical equipment regularly.

The operator should make a quick visual survey when first arriving at the filter unit. All motors and mechanical devices should be operating normally. Where an abnormal condition is indicated, prompt corrective actions, in the form of adjustments, lubrication, or repairs, may be necessary. When operation seems normal, the operator should then inspect and lubricate the mechanical devices, according to the manufacturer's maintenance instructions. Effluent quality tests and housekeeping tasks will complete the routine operation duties. In addition to a record of test results, a log of the filter operation (including a written record of all repairs, adjustments, lubrications, and unusual operating conditions) is strongly recommended.

*P*OLISHING POND

Polishing ponds are an additional form of treatment used in situations where impermeable soils and sufficient area for construction are available. They are simple to operate because they are nonmechanical and can produce an effluent low in BOD. In areas where land costs are inexpensive, they can be the most cost-effective method of additional treatment. Figure 10.4 illustrates how the basic function of a polishing pond occurs.

There must be an active invertebrate population for the pond to polish the effluent. The most efficient aerobic scavengers include aquatic snails, crus-

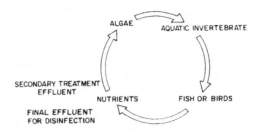

Figure 10.4   Basic function of a polishing pond.

Ed: and? of? instead of or various

sp? dameselfly

taceans (daphnia and crayfish), coleopteran (water beetles), water boatmen, and the aquatic instars (stages between molts), or various insects, such as the dragonfly and dameselfly. Aquatic snails are nature's vacuum cleaners, consuming slimes (made up principally of particles of waste, single-cell bacteria, and algae) and zooplankton that collect on underwater surfaces. Zooplankton are small aquatic animals (amphipods and ostracods) that improve water quality by consuming particles of algae, protozoa, colloids, and bacteria. Daphnia are filter feeders which, in addition to feeding on algae, also consume other particulate matter suspended in the pond water.

Water beetles and water boatmen are two of the most common pond insects. The predaceous diving beetle feeds on worms and dead organisms, whereas the water strider eats anything that lands on the surface of the water, insect, or vegetation.

The dragonfly is a voracious eater from the time it is a small aquatic nymph to its adulthood. As a nymph, no mosquito larvae or other aquatic insect can escape the extendable lower lip of the dragonfly. Dragonflies are beneficial because they continually destroy such pests as mosquitoes, flies, and other winged insects.

Aerobic scavengers need dissolved oxygen and circulation to remain healthy and able to reproduce. This condition can be maintained with various floating aerators that are electric and wind-powered. Air lift systems also provide a good combination of circulation and aeration; however, in most instances, no mechanical aeration is used.

There will almost always be plenty of oxygen during the day if there are algae in the pond. If the dissolved oxygen level falls below 1 mg/L (1 ppm) just before dawn, more aeration is needed. Powered aerators should be operated for 0.5 h at 3-h intervals during the day, and continuously from sundown to 2 h after sunrise.

A *polishing pond* is generally designed to retain approximately 10 days of design flow from the package plant. This provides adequate time for the natural biological actions of photosynthesis and bacterial degradation to further treat the wastewater. The pond should have a level bottom with embankments sloped at 1:3 to 1:6 to prevent weed growth and allow for mowing; its depth will be between 1 and 2 m (3 and 6 ft). The pond should be enclosed with a fence that has a locked entrance gate; the fence and gate should be grounded

---

electrically as a safety precaution. The discharge pipe must have a tee or ell that draws the effluent from approximately 0.5 m (18 in.) below the water surface; additional design specifications may vary.

Maintenance will involve care of the fence and embankments and weed control. Erosion should be controlled by seeding any bare areas on the embankments. Large stone or concrete riprap at the water's edge is also beneficial. Burrowing animals, including muskrats and turtles, must be removed before serious damage causes a slip that can lead to dike failure. The grass should be cut regularly during the warmer months, especially at the waters edge; all grass clippings must be kept out of the pond.

One common maintenance problem for polishing ponds is the overgrowth of duckweed or floating algae. Either plant can completely cover the surface of the pond quite rapidly, blocking sunlight from penetrating the water and causing the pond to become septic. **As a last resort,** both plants can be removed by spraying with special herbicides. Before choosing an herbicide, the operator should be sure it is labeled as effective against the specific plant causing the problem. Any type of chemical control will interfere with the natural treatment occurring in the pond. Chemicals kill plants (unlike harvesting, where the plants are physically removed); consequently, the dead plants decay and provide additional nutrients for new and more vegetation.

Cattails and other rooted vegetation must be removed from the edge of the pond before they become well-established. These rooted plants prevent thorough mixing, contribute to mosquito breeding problems, and may ruin the pond seal.

Another common problem with small polishing ponds is that their effluent is high in pH and suspended solids (mostly algae) during warmer months, resulting in permit violations with respect to these parameters. In many instances, secondary treatment plant effluent is higher in quality than effluent from the polishing pond. A design detention time of less than seven days may prevent algae growth. Further information on the operation of polishing ponds is given in a U.S. Environmental Protection Agency (U.S. EPA) operations manual entitled *Municipal Wastewater Stabilization Ponds* (U.S. EPA, 1983).

*P*HOSPHORUS REMOVAL

Phosphorus is one of the most important nutrients for plant growth. Because secondary treatment plants do not remove a significant amount of phosphorus, the effluent discharged can cause large amounts of algae to grow in the receiving water, especially lakes. As a preventative step, some treatment plants are required to remove phosphorus by an additional treatment step (sometimes referred to as *tertiary treatment*).

The most common method for phosphorus removal is to add lime [$Ca(OH_2)$], alum [$Al_2 (SO_4)_3 \cdot 14H_2O$], or ferric chloride ($FeCl_3$) to the wastewater. In each case, the calcium, aluminum, or iron reacts chemically with the phosphorus to form solid particles that can be removed by settling. Sometimes a coagulant aid, such as a polyelectrolyte, can be added to improve the removal of suspended particles.

Perhaps the most difficult question the operator faces is determining how much chemical should be added. Enough chemical must be added to ensure the process is complete, yet adding extra chemicals is uneconomical and may even make the process less efficient. Therefore, the amount of chemical to be added is related to the amount of phosphorus in the wastewater. Domestic wastewater contains 20 mg/L soluble phosphate; such a system requires 60 mg/L of commercial alum to precipitate the phosphorus. For further details consult the Water Environment Federation's MOP 11, *Operation of Municipal Wastewater Treatment Plants* (1998).

An operator attending to an advanced waste treatment process must devote a great deal of time and attention to ensure successful operation. The operator must check the chemical pumps daily and inspect tube plugging and solution levels; all information must be entered into the plant log. The flash mixer should be checked for proper operation and the settling tank checked for sludge depth. The alert operator will recognize that adding chemicals creates more sludge and necessitates more frequent sludge removal than in secondary plants without phosphorus removal.

Finally, if phosphorus removal is discontinued for a prolonged period, the chemical pumps and lines should be flushed with clean water to prevent corrosion problems.

*P*OSTAERATION

Often, effluent leaving the package plant has a low dissolved oxygen content (less than 2 mg/L) that must be increased before discharge. The simplest method is to allow the effluent to flow over a series of steps before it enters the receiving stream. Typically, the only maintenance required is to occasionally clean the steps; yet, in some cases, where there is insufficient elevation drop, mechanical or diffused aeration becomes necessary. In either case, operator attention is required to ensure satisfactory performance of the equipment. The measurement of dissolved oxygen in the effluent can best indicate satisfactory performance.

*N*ITROGEN REMOVAL

Nitrogen is also an important plant nutrient, because high concentrations in the form of ammonia (NH_3) can be toxic to fish. Furthermore, organic nitrogen and ammonia are eventually converted to nitrate nitrogen (NO_3). Because this process requires 2.1 kg (4.6 lb) of oxygen for every kilogram (pound) of ammonia converted, oxygen depletion can occur in the receiving stream if the conversion is not done in the treatment plant. Fortunately, most package treatment plants use the extended aeration process, where nitrification can take place under favorable conditions in the treatment plant, according to the following reaction.

$$2\ NH_3 + 4\ O_2 \rightarrow 2\ NO_3 + 2H^+ + 2\ H_2O \qquad (1)$$
(ammonia) + (oxygen) (nitrate) + (hydrogen) + (water)

Because the nitrification reaction is very sensitive to pH, the operator should maintain pH between 6.5 and 8.5.

While nitrification can handle potential oxygen depletion problems, nitrogen in the effluent is still available for plant growth. Nitrogen removal in the treatment plant can be accomplished by denitrification. As shown in eq 2, nitrate is converted to nitrogen gas, which is lost to the atmosphere.

$$NO_3 + [\text{carbon compound}] \rightarrow N_2 + CO_2 + H_2O + OH^- \qquad (2)$$
(nitrate) (nitrogen gas) (carbon dioxide gas) (water) (hydroxide)

Because this chemical reaction can occur only under anaerobic conditions, it typically takes place in a separate tank. Methanol was used widely in the past as the primary carbon source, but more recent techniques have allowed wastewater to serve as the carbon source. Similar to other advanced waste treatment processes, denitrification requires close operator attention.

Operators of secondary treatment plants may associate denitrification with rising sludge. When activated sludge remains in the clarifier too long, anaerobic conditions can induce denitrification and cause the nitrogen gas to carry sludge particles to the surface.

REFERENCES

Rich, L. G. (1980) *Low Maintenance Mechanically Simple Wastewater Treatment Systems*; McGraw-Hill: New York.

U.S. EPA (1983) *Municipal Wastewater Stabilization Ponds*, 625/1-83-015. U.S. Environmental Protection Agency: Cincinnati, Ohio.

Water Environment Federation (1998) *Operation of Municipal Wastewater Treatment Plants*, 5th ed. (MOP No. 11); Water Environment Federation: Alexandria, Virginia.

Chapter 11
Corrosion Control and Climate Conditions

*I*NTRODUCTION

Why do metals and concrete corrode? There are two basic causes of corrosion; one is chemical attack and the other involves electrochemical reactions between metals and their environment. This chapter reviews the causes of corrosion that affect package plants and makes recommendations to prevent corrosion and ensure that package plants will continue to provide long-term wastewater treatment services for the owner.

*C*HEMICAL ATTACK

Direct chemical attack is associated with attacks on metals by acids, alkalies, and other similar corrosive chemicals. This contributor to corrosion in wastewater treatment plants can occur where chemicals such as alkalies or acids are used for pH control, or where chlorine is used for disinfection of wastewater before discharge. This type of corrosion can be minimized greatly if materials resistant to chemical attack are used. Plastics and fiberglass materials are available for piping or tubing, chemical storage tanks, and pumping equipment used for chemical feed systems in wastewater treatment plants. These materials are typically incorporated to the treatment plant design and are installed at the time the plant is constructed. Older treatment plants, however, may have been built with other materials that may have been worn or corroded over time by direct chemical attack. When any of these items must be replaced, plastic or fiberglass materials suited for the intended use should be considered. The required items and components such as tubing or pipe, chemical feed pumps, and tanks can typically be obtained from plumbing houses or equipment manufactures. Care should be used in the selection of such items to ensure they will resist corrosive attack by the chemicals with which they may come in contact.

*E*LECTROCHEMICAL *CORROSION*

The second, and more common, type of corrosion is known as electrochemical (galvanic) corrosion, and is most obvious in the case where two dissimilar metals are electrically coupled. A battery is an excellent example of this corrosion process. One metal, acting as an anode, provides electrons to the second metal. As a result of providing these electrons, the single metal dissolves into the solution in contact with both metals. This action causes metal loss and eventually destroys the device supplying the electrons. Figure 11.1 illustrates this action. The solution carrying the metal ions is known as an electrolyte. The electrons generated as a result of this reaction travel to the cathode, where a second reaction occurs. This electrolytic type of corrosion occurs wherever two dissimilar metals are in contact, or when both are in contact with an electrolytic solution. This electrolytic solution may be chemicals stored in tanks or chemical feed tanks, wastewater in any one of several treatment processes, or a thin liquid film coating the various metal components.

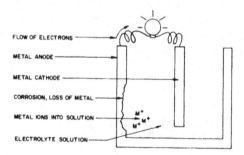

Figure 11.1   Electrochemical (galvanic) corrosion.

There can be numerous opportunities for this type of corrosion of waste-water treatment plants because metals are used for such items as piping, tanks, grating, pumps, and blowers. Awareness of this type of electrochemical corrosion may alert the operator to a potential situation where corrosion may occur. The most effective preventive methods are to use similar metals that will be in contact with each other, isolating the two metal components with a nonmetallic material such as fiberglass or plastic, or to completely substitute the metallic components with nonmetallic ones. Although there may be a situation where dissimilar metals could corrode, it is important to note that this process may occur very slowly. Therefore, the operator should occasionally check these locations for evidence of corrosive attack.

Metal corrosion can also occur when only one metal is in contact with an electroytic solution. These electrolytic cells are called *differential aeration cells* and occur when there are different concentrations of oxygen. As indicated in Figure 11.2, this corrosion can develop when iron rusts. The concentration of oxygen under the rust is lower than that around the outside of the iron and rust, thereby resulting in corrosion. The threaded surfaces of two coupled pipes provides a suitable environment for this type of corrosion. This corrosion can account for the pitting damage that can occur under rust or at the water-air interface of the tank wall.

Corrosion can also occur in metallic components such as piping, tanks, door and window frames, and grating, where there is no contact with a dissimilar metal and where oxygen concentrations do not vary from one metallic surface to another. The process by which this type of corrosion occurs is electrochemical in nature. However, metals are not totally pure and homogeneous, but contain small quantities of impurities; these impurities can act as anodes or cathodes with the primary metal. This is especially true of iron and steel, where the anodes and cathodes cannot be isolated. The metal surface must receive a protective coating to prevent contact with an electrolytic solution or the atmosphere. Although there are several other mechanisms that can cause corrosion, those just described are generally associated with corrosion in wastewater treatment plants.

In addition to the preventive methods already mentioned, there are other ways to minimize corrosion. For example, the environment may contribute to the corrosion reaction. Treatment plants enclosed within buildings may cause damp conditions; this moisture, in combination with particulate matter in the atmosphere, can produce the proper electrolyte to initiate corrosion. Proper ventilation can greatly reduce corrosion potential by removing or reducing

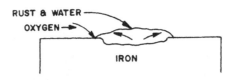

Figure 11.2 Corrosion by differential aeration cells.

this electrolyte source. Other methods to control or increase corrosion resistance of metal include use of protective metal coatings, production of oxides or phosphates for coating iron and steel, application of protective paints, and other surface treatments.

All of these methods (with the exception of paint application) require specialized skill, knowledge, or equipment for their application. Corrosion protection for cathodic or anodic protection is often used for large structures, such as buried or submerged tanks.

PROTECTIVE COATING SYSTEMS

Typically, the only practical method the plant operator can use to control and minimize corrosion is to apply protective paint. There are numerous paint types available for use as protective coatings. The type selected will be determined by the surface to be coated, the type of atmosphere to which it is or will be exposed, and any desired special paint characteristics, such as electrical insulation or heat reflection.

The development of a standard coating system will be very useful to maintain a package plant. An example protective coating system is shown on Table 11.1. Another useful tool is a standard color schedule as shown on the example Table 11.2.

BUILDINGS

Small buildings that are a part of package plants should be painted with a high-quality interior or exterior paint, as appropriate. Keep in mind these buildings are exposed to the same corrosive environment as the treatment plant itself. There are many types of high-quality epoxy and latex paints available. Color should be whatever presents a pleasing appearance, such as light shades of brown and tans.

PIPING

Steel piping that is submerged, such as is in a clarifier, can be coated with a coal tar epoxy coating system, which provides at least two protections. Like most coating systems, coal tar epoxy isolates the pipe from the corrosive environment of wastewater. Second, because coal tar epoxy coating systems are thick, they provide protection against abrasion from other equipment, operator equipment and tools, and other things.

Table 11.1 Example protective coating system.

Coating system number	Coating material	Surface types	Surface preparation
No. 1 3-coat urethane	Primer Polyamide epoxy (4 mils[a] DFT) Intermediate High solids epoxy (6 mils DFT) Finish Acrylic polyurethane enamel (2.5 mils DFT)	Metal (steel) Severe exposure environment, non-submerged exterior, such as clarifier catwalks, tanks, etc.	SSPC SP[b]-6 (Commercial blast cleaning)
No. 4 Coat tar epoxy	Amine cured or polyamide Cured epoxy (16 to 20 mils DFT)	Metal (steel) Submerged environment, such as clarifier components Concrete Submerged environment, such as tank walls	SSPC SP-10 (Near white blast cleaning) Brush blast
No. 8 High solids epoxy	Amine cured epoxy (5 to 7 mils DFT)	Metal (steel) Non-submerged interior metal such as piping, fittings, tanks, pumps, etc.	SSPC SP-1 (Solvent cleaning) SSPC SP-2 (Hand tool cleaning) SSPC SP-3 (Power tool cleaning)
No. 9 2-Coat urethane epoxy	Primer High solids epoxy (6 mils DFT) Finish Acrylic polyurethane enamel (2.5 mils DFT)	Metal (steel) Severe exposure, exterior sunlight exposed, such as doors and piping	SSPC SP-1 (Solvent cleaning) SSPC-2 (Hand tool cleaning) SSPC-3 (Hand tool cleaning)

[a] mils $\times$ 25.40 = μm
[b] SSPC SP = Society for Protective Coatings Surface Preparation (number indicates scale).

Table 11.2   Example coating system color schedule.

Description	Color
Raw wastewater pumps and piping	Brown
Sludge/grit equipment, pumps, and piping	Brown
Effluent pumps and piping	OSHA[a] safety green
Electrical enclosures (remember that some of this equipment may be owned and maintained by the local power company and as such should be coated by them)	OSHA safety green
Digester gas, compressors, and piping	OSHA safety yellow
Paintable grating and walkways	Same color as associated equipment (aeration basin walkway should be brown)
Potable water	OSHA safety blue (property label to avoid confusion)
Boilers	OSHA safety orange

[a]Occupational Safety and Health Administration (OSHA) colors are suggested because normally they have a glossy finish that results in a very pleasing appearance.

CONCRETE

Concrete that is not exposed to moisture or submersion can be treated similar as to a building and coated with suitable protective coatings. Concrete that is in a submerged environment, such as tank walls, should be coated with a coating system similar to coal tar epoxy, as described above for steel piping.

STEEL

Because steel (with the exception of steel alloys, such as stainless steel) is not highly corrosion-resistant, treatment plant components manufactured from steel will require a protective coating. When a treatment plant is constructed, exterior and interior steel surfaces should receive proper corrosion protection. In addition, some steel components, such as buried tanks, may be provided cathodic protection through a sacrificial magnesium anode. Nonetheless, steel surfaces and components may begin to experience corrosion over time. This can occur when the effectiveness of paint or other coating decreases under repeated attacks by moisture, fungus, sunlight, or age. In addition, nicks in painted steel surfaces expose base metal, providing an opportunity for corrosion.

Major painting of large buried tanks will be beyond the capability of even large, fully staffed treatment facilities. When such a situation arises, the operator is advised to seek professional assistance. This may involve preparing specifications and advertising for contractors to perform the required work. Less extensive painting requirements may permit the operator to perform these paint tasks. Care should be taken to prevent paint from getting into the wastewater. Components to be painted should be evaluated for any possible signs of corrosion. This evaluation may enable the operator to apply other corrective measures that may extend the life of the paint or reduce the severity of potential corrosion. Steel and iron surfaces to be painted should be thoroughly cleaned first. Solvents will aid in removing oil and grease, while scale and rust may be removed by scraping, power wire brushing, power sanding, or sandblasting. A commercial phosphoric acid wash is recommended before painting, unless the surface has been sandblasted. The phosphoric acid wash aids in bonding the paint and provides a thin phosphate coating on the steel surfaces to retard corrosion. This procedure is especially important in the case where an existing paint film has been scratched.

Steel surfaces should receive a primer after cleaning and acid washing before painting. The primer may be applied in one or more coats and should contain rust-inhibiting pigments. This will also provide a bonding surface for the top coat of paint. Paint suppliers can generally be helpful in selecting the proper coatings for he proposed application. Where hydrogen sulfide exposure is possible, lead-base paints should not be used, as hydrogen sulfide will cause the lead to break down.

GALVANIZED IRON/STEEL

Typically, galvanized iron will better resist corrosion found in wastewater treatment plants. Yet, when the coating becomes nicked or scratched, the potential for corrosion is increased. Galvanized iron should be treated with a proprietary etching solution or phosphoric acid wash before priming and painting. The paint used for the prime coat should contain zinc chromate. **Primers containing red lead should not be used** because of the possible galvanic or electrochemical reaction between the lead in the primer and the zinc; furthermore, lead paints are toxic.

ALUMINUM

As with the applications of all other coatings, surface preparation for aluminum is important. This includes removal of any grease and oil. The aluminum that is to be painted must first be treated with a phosphoric acid wash, because paints do not typically adhere well to aluminum. A prime coat containing zinc chromate should be applied before the top coat. This will prevent a similar galvanic reaction, mentioned previously for galvanized iron.

Magnesium Anodes

For metal plants, in addition to painting, periodic replacement of magnesium anodes is required. Frequency of replacement will depend on the painting of the exterior surface and soil conditions. Because placement of the magnesium anode sacks makes them difficult to inspect, it is recommended that they be replaced every 10 years. A location sketch should be made when the plant is installed.

Climatic Conditions

PIPE AND EQUIPMENT INSULATION. The following is a general guideline for materials that can be used for insulating piping materials. This section specifies insulation for exposed piping and related equipment and appurtenant surfaces. Low-temperature-class insulation should be suitable for an operating temperature range of −73.33 to 37.78°C (−100 to +100°F).

MATERIALS. Piping insulation should be tubular. Insulation for valves, fittings, expansion joints, flanges, and other connections should be segmented sections, molded, or blanket-type coverings of the specified type and thickness of pipe insulation. Equipment insulation should be blanket or rigid board-type cut to fit the surface. Low-temperature-class insulation should be of the unicellular elastomeric thermal, cellular glass, or fiberglass type.

Unicellular elastomeric thermal-type insulation should conform to the requirements of ASTM C534, Type 1. Cellular glass type insulation should conform to the requirements of ASTM C552, Type ll.

Fiberglass-type insulation should conform to the requirements of FEDSPEC HH-1-558B. Calcium silicate-type insulation should conform to the requirements of ASTM C553, Type ll, Class C. Laminate jackets should consist of aluminum and white kraft paper. Jackets should have a perm rating for water vapor transmission of not more than 0.02 in accordance with procedure A of ASTM E96. Aluminum jackets should be constructed of smooth finish aluminum sheet conforming to ASTM B209, alloy 5005, temper H16, with integral vapor barrier. Jackets should be 0.04 cm (0.016 in.) thick. Sheet metal screws should be aluminum or stainless steel. Jackets should be secured with 0.05×1.9-cm (0.020×0.75-in.) type 304 stainless steel expansion bands. Polyvinylchloride covers should be one piece, per-molded polyvinylchloride conforming to FEDSPEC L-P1535E, composition A, Type ll, Grade E4. Aluminum covers should be constructed of smooth finish aluminum sheet conforming to ASTM B209, alloy 5005, temper H16 with integral vapor barrier. Covers should be 0.04 cm (0.016 in.) thick. Shields should be 16-gauge galvanized steel sheet, formed into a half cylinder. Shield length should be as recommended by the insulation manufacturer. Flashing should include aluminum caps, sealant, and reinforcing. Aluminum caps should be 20-gauge thick and should be cut to completely cover the insulation. Sealants should be as recommended by the insulation manufacturer. Reinforcement in flashing heated up to 187.78°C (370°F) should be nylon fabric. Reinforcement in flashing for hotter surfaces should be wire mesh or recommended by the insulation manufacturer.

INSTALLATION. Insulation should be applied over clean, dry surfaces. Double-layer insulation, where specified, should be provided with staggered section joints. Hangers, anchors, pipe guides, and other support elements should not interrupt jackets, covers, and insulation. Insulation and jackets should be protected from crushing, denting, and similar damage during construction. Vapor barriers should not be penetrated or otherwise damaged. Insulation, jacket, and vapor barriers damaged during construction should be removed and new material should be installed. Unless otherwise specified, piping insulation should be provided with laminated jackets. Insulation should be butted firmly together and jacket laps and joint strips provided with lap adhesive. Jackets should be provided with their seams located on the underside of pipe. Fitting, connection, flange, and valve insulation should be provided with a rigid cover. Insulation should be secured in place with 20-gauge wire and a coat of insulating cement. Covers should overlap the adjoining pipe insulation and jackets. Covers should be provided with their seams located on the underside of fittings and valves.

Low-temperature-class insulation should have ends sealed off with a vapor barrier coating. For pipe sizes approximately 5 cm (2 in.) and smaller, insulation should be provided with rigid polyvinyl chloride (PVC) covers. Covers should be sealed at edges with vapor barrier adhesive. The ends of covers should be secured with vinyl tape. The tape should overlap the jacket and the cover at least 2.54 cm (1 in.). The vapor barrier should not be penetrated. For pipes 6.35 cm (2.5 in.) and larger, insulation should be provided with rigid aluminum covers. Covers should be mechanically secured by corrosion-resistant tacks pushed into the overlapping throat joint. Unless otherwise specified, metal shields should be provided at pipe supports. The inside face of each shield should be coated with insulation adhesive to prevent movement. Additional support should be provided at each shield as specified on the drawing details.

Flashing should be provided at jacket penetrations and terminations. Clearance for flashing should be provided between insulation system and piping supports. A heavy tack coat of sealant should be troweled over the insulation, extending over the jacket edge 2.54 cm (1 in.) and over the pipe or protrusion approximately 5 cm (2 in.). Reinforcement should be stretched over the tack coat after clipping to fit over pipe and jacket. Clipped reinforcing should be strapped with a continuous band of reinforcing to prevent curling. Sealant should then be troweled over the reinforcement to a minimum thickness of 0.3175 cm (0.125 [1/8] in.).

Aluminum caps should be formed to fit over the adjacent jacketing and completely cover coated insulation. A cap should be held in place with a jacket strap.

WINTER PROOFING

This section addresses freeze protection throughout the plant. The following is to be used as a guideline and is not inclusive of all steps necessary to protect the plant from cold weather.

SAFETY

Remember that cold weather brings on many problems. Standing water freezes and becomes slippery. Extension cords are trip hazards, and so on. When preparing the plant for winter, always keep an eye out for potential safety hazards. The following is a general checklist for safety.

(1) Remove hose from the bibs and store inside. Attach short hoses to the bibs.
(2) Ensure that interior unit heaters are on and working.
(3) If a unit is out of service, leave the drain open and turn on the sprayers if water cannot be isolated and completely drained out of the sprayers.
(4) Ensure heat tape on piping and equipment is plugged in and working.
(5) Wrap and insulate water lines if necessary.
(6) Plug in light bulbs used for heating.
(7) If a clarifier is out of service, remove the cap from the spray line and allow water to run into the launder.
(8) Ensure liquid lines are drained.
(9) Turn on heater in the auto sampler.
(10) **Open all valves** on out-of-service units.

SUMMARY

Corrosion prevention and correction is a continuous, ongoing process in wastewater treatment plants. The operator must be alert to conditions within the treatment plant that enhance corrosion reactions. Equally important is proper treatment of corrosion once it is discovered. Preventing or minimizing some types of corrosion will extend the life of corrosion prevention measures when corrosion is encountered. If corrosion recurs frequently, substitute materials should be considered. The use of protective coatings requires proper surface preparation and selection of the primer/paint system to resist potential corrosive atmospheres. Where extensive repainting is necessary, preparing specifications and advertising for experienced contractors should be considered. Other protective measures, such as anodic or cathodic protection, may be desirable. The operator should retain the services of a qualified engineer when extensive repainting or other corrosion prevention measures are to be used. Remember that

- Corrosion of package plants by wastewater is a natural process that must be prevented or at least arrested.
- Generally, new package plants are manufactured with corrosion protection features and the owner, contractor, or engineer should work with the manufacturer to ensure the protection system is understood, not compromised, and completed correctly.
- Owners and operators of existing package plants need to realize there is a tremendous amount of information and help available on corrosion protection systems. Sources of help include the original plant manufacturer and industrial coating manufacturers, suppliers, contractors, and engineers.
- Industrial coatings must be applied safely and generally need to be applied by professional, experienced, licensed, industrial coating contractors. Modern coating systems can be very dangerous to one's health if improperly applied.
- Coating systems should be kept simple and stockpiling excess spare supplies avoided.

DEFINITIONS AND ABBREVIATIONS

TOLERANCES. The numerical information quoted in the product datasheets have been derived from laboratory test data obtained under controlled conditions for the products described. While every effort has been made to ensure accuracy, this information will be subject to minor variations obtained in normal manufacturing tolerances, and any fluctuations in ambient conditions during the application and curing periods.

DRY FILM THICKNESS. The measured thickness of the final dried film applied to the substrate.

WET FILM THICKNESS. The initial thickness of the wet coating applied to the substrate.

VOLUME SOLIDS. The volume solids figure given on the product data sheet is the percentage of the wet film, which remains as the dry film, and is obtained from a given wet film thickness under specified application method and conditions. These figures have been determined under laboratory conditions using the test method described in the Oil and Colour Chemists (OCCA) Monograph No. 4, *Determination of the Solid Content of Paint (by Volume)*. This method is a modification of ASTM D-2697, which determines the volume solids of a coating using the recommended dry film thickness of the coating quoted on the product data sheet and a specified drying schedule at ambient temperature, i.e., 7 days at 23°C.

DRYING TIME. The drying times quoted in the product data sheet have been determined in the laboratory using a typical dry film thickness, ambient temperature quoted in the relevant product data sheet, and appropriate test method.

The drying times achieved in practice may show some slight fluctuation, particularly in climatic conditions where the substrate temperature differs significantly from the ambient air temperature.

OVERCOATING INTERVAL. The product data sheet gives both a minimum and maximum overcoating interval, and the figures quoted at the various temperatures are intended as guidelines, consistent with good painting practices.

FLASH POINT. The flash point is the minimum temperature at which a product, when confined in a Setaflash closed cup, must be heated for the vapors emitted to ignite momentarily in the presence of a flame (ISO 3679:1983)

VOLATILE ORGANIC CONTENT. Volatile organic content (VOC) is the weight of organic solvent per liter of paint.

U.S. ENVIRONMENTAL PROTECTION AGENCY FEDERAL REFERENCE METHOD 24. The U.S. Environmental Protection Agency (U.S. EPA) published procedures for demonstration of compliance with VOC limits under Federal Reference Method 24—*The Determination of Volatile Matter Content, Density, Volume Solids and Weight Solids of Surface Coatings.* This method was originally published in the Federal register in October 1980 and coded 40 CFR, Part 60, Appendix A, and amended in 1992 to incorporate instructions for detailing with multicomponent systems and a procedure for the quantitative determination of VOC-exempt solvent.

It is recommended that users check with local agencies for details of current VOC regulations to ensure compliance with any local legislative requirements when proposing the use of any coating.

WORK POT LIFE. The work pot life is the maximum time during which the product supplied as separate components should be used after they have been mixed together at the specified temperature (ISO 9514:1922). The values quoted have been obtained from a combination of laboratory tests and application trials, and refer to the time periods under which satisfactory coating performance will be achieved.

Application of any product after the working pot life has been exceeded will lead to inferior product performance, and must **not** be attempted, even if the material in question appears as liquid in the can.

SHELF LIFE. The shelf life quoted on the product data sheets is generally a conservative value, and it is probable that the coating can be applied without any deterioration in performance after this period has elapsed. However, if the specified shelf life has been exceeded, it is recommended that the condition of the material is checked before any large scale application is undertaken using materials beyond the quoted shelf life.

REFERENCES

American Society for Testing and Materials (2003) Standard Specification for Aluminum and Aluminum-Alloy Sheet and Plate, ASTM B209, West Conshohocken, Pennsylvania.

American Society for Testing and Materials (2003) Standard Specification for Mineral Fiber Blanket Thermal Insulation for Commercial and Industrial Applications, ASTM C553, Type ll, Class C, West Conshohocken, Pennsylvania.

American Society for Testing and Materials (2003) Standard Specification for Preformed Flexible Elastomeric Cellular Thermal Insulation in Sheet and Tubular Form, ASTM C534, Type 1, West Conshohocken, Pennsylvania.

American Society for Testing and Materials (2003) Standard Specification for Cellular Glass Thermal Insulation, ASTM C552, Type ll, West Conshohocken, Pennsylvania.

American Society for Testing and Materials (2003) Standard Test Method for Volume Nonvolatile Matter in Clear or Pigmented Coatings, ASTM D-2697, West Conshohocken, Pennsylvania.

American Society for Testing and Materials (2003) Standard Test Methods for Water Vapor Transmission of Materials, ASTM E96, Procedure A, West Conshohocken, Pennsylvania.

FEDSPEC L-P1535E, composition A, Type ll, Grade E4.

Federal Reference Method 24—*The Determination of Volatile Matter Content, Density, Volume Solids and Weight Solids of Surface Coatings.*

Oil and Colour Chemists (OCCA) Monograph No. 4, *Determination of the Solid Content of Paint (by Volume).*

Chapter 12
Regulatory Requirements

INTRODUCTION

Regulatory controls for package plants will vary widely, depending on the receiving water body, geographic location, and permitting authority. Plants in the United States, which discharge to a receiving water body, will be regulated by a permit issued by either the U.S. Environmental Protection Agency (U.S. EPA) or a state, which has received the delegated authority from U.S. EPA. Plants located in Canada will be regulated by the province in which they are located.

NATIONAL POLLUTANT DISCHARGE ELIMINATION SYSTEM PERMITS

Section 402 of the U.S. Federal Clean Water Act (U.S. EPA, 2002a) requires the issuance of a National Pollutant Discharge Elimination System (NPDES) permit for the discharge of any pollutant, or combination of pollutants, into the waters of the United States. It also requires all publicly owned treatment works (POTWs) to meet secondary effluent limitations or, where necessary, more stringent effluent limitations to meet water quality standards. For those plants that are not POTWs (not owned by a city, county, or authority-type government), the Clean Water Act requires that effluent limitations shall require the application of the best practicable control currently available, which most state regulatory agencies consider to be more stringent than secondary effluent limitations.

EFFLUENT LIMITATIONS

Secondary effluent limitations are defined in the Code of Federal Regulations, Title 40, Part 133 (40 CFR Part 133) (U.S. EPA, 2000b) as follows: for bio-chemical oxygen demand (BOD) and suspended solids, the 30-day average shall not exceed 30 mg/L, and the 7-day average shall not exceed 45 mg/L. In addition, the 30-day average percent removal shall not be less than 85% (Note: At the option of the permitting authority, carbonaceous biochemical oxygen demand [CBOD] may be substituted in lieu of BOD). In this case, the 30-day average shall not exceed 25 mg/L and the 7-day average shall not exceed 40 mg/L. The primary purpose for this is to compensate for the nitrification component, which adds to the perceived BOD value. This may be extremely critical to facilities which are required to nitrify to meet water quality limitations. For pH, the effluent value shall not exceed 6.0 to 9.0 standard units, unless the POTW can demonstrate that an excursion is not caused by industrial wastes or by the addition of inorganic chemicals.

Section 301 (b) of the Clean Water Act (U.S. EPA, 2002b) requires that a POTW must meet more stringent effluent limitations than secondary limitations when necessary to meet water quality standards. The 40 CFR Part 122.44 (U.S. EPA, 2000a) requires the permitting authority to establish effluent limitations for all pollutants that cause or may cause a violation of any water quality standard, including a narrative water quality standard. For most extended aeration package plants, the parameters most likely to be regulated are ammonia-nitrogen and dissolved oxygen. These, in addition to BOD, are addressed for the purpose of maintaining minimum dissolved oxygen levels in the receiving waters. However, a number of other parameters are now being regulated, which include, but are not limited to, phosphorus, chlorine residual, metals, and toxicity. These items will not be addressed other than to state that if they are included in the NPDES permit, the operator should consult with the permitting authority to ensure proper compliance. The operator should carefully review their permit as most NPDES permits will contain generic boilerplate language regarding toxicity in the wastewater discharge.

SAMPLING AND ANALYSIS

The 40 CFR 133.104 (U.S. EPA, 2000b) requires that sampling and analysis procedures for regulated pollutants shall be in accordance with guidelines listed in 40 CFR Part 136 (U.S. EPA, 2000c), which contains the approved test procedures, appropriate sample containers, and sample preservation and holding times. It is recommended that the operator check with the local regulatory agency for any additions or modifications to sampling and analysis procedures. For most operators, the test methods that they will probably use will be found in the current edition of *Standard Methods for the Examination of Water and Wastewater* (APHA et al., 1998). Table 12.1 summarizes requirements for most parameters that the operator will encounter.

INTRODUCTION OF POLLUTANTS AND TOXICITY

For plants that have a service area, which has commercial and/or industrial discharges, the operator must pay particular attention to the introduction or pollutants from nondomestic sources. Some pollutants may cause plant upsets, which may lead to permit violations, and others may pass through without affect to the plant, but which again may lead to a permit and/or water quality violation. The best way to avoid a problem from nondomestic pollutants is to practice pollution prevention and know the plant's customers. It is easier to keep some pollutants our of the sewerage system than to treat them or address any other problems they may cause at the plant. Depending on the type of discharge, it may be necessary to report, per Section 301 or 306 of the

Table 12.1 Parameters for typical sampling requirements.

Parameter	Standard method (APHA et al., 1988)	Container	Preservative	Holding time
BOD$_5$	5210	Plastic, glass	Cool, 4 °C	48 hours
TSS	2540-D	Plastic, glass	Cool, 4 °C	7 days
Ammonia– nitrogen	4500-NH$_3$	Plastic, glass	H$_2$SO$_4$ to pH <2, Cool, 4 °C	28 days
Chlorine residual	4500-Cl	Plastic, glass	None	Analyze immediately
Fecal coliform	9221-E 9222-D,E	Plastic, glass	Cool, 4 °C	6 hours
pH	4500-H	Plastic, glass	Cool, 4 °C	Analyze immediately

Clean Water Act (U.S. EPA, 2002b), to the permitting authority. It is recommended that the operator discuss a proposed nondomestic discharge with their permitting authority before any such discharge is allowed to connect to the sewerage system.

BIOSOLIDS

The U.S. EPA promulgated 40 CFR Part 503 regulations on February 19, 1993 (U.S. EPA, 2000d). The 503 regulations, also known as the *Standards for the Use or Disposal of Sewage Sludge*, established standards that consist of general requirements, pollutant limits, management practices, and operational standards, for the final use or disposal of wastewater sludge generated during the treatment of domestic wastewater in a treatment works. These standards included wastewater sludge applied to the land, placed on a surface disposal site, or disposed in an incinerator. The standards also included requirements for monitoring and record keeping. It applies only to wastewater sludge applied to the land, placed on a surface disposal site, or disposed in an incinerator. Sludge that is hauled to a sanitary landfill for codisposal with other wastes in the landfill is not regulated by the 503 regulations. Most package plants will probably use this method of disposal.

However, sludge generated from a treatment works treating domestic wastewater may be beneficially used. These solids are nutrient-rich organic materials that will provide nitrogen, phosphorus, and trace amounts of micronutrients. The term *biosolids* refers to the treated wastewater sludge, which meets the pollutant concentration limits, pathogen reduction, and vector attraction reduction requirements found in federal regulations 40 CFR Part 503 (U.S. EPA, 2000d). The U.S. EPA 503 regulations encourage the beneficial use of biosolids under controlled conditions. However, even digested or chemically treated sludge may contain pathogens or metals that could potentially harm plants, animals, or humans if applied at too great a rate. Therefore, the objective of these guidelines is to allow for the use of the nutrients and other beneficial properties of biosolids, while ensuring that the health and welfare of the population is protected.

Most NPDES permits will require modification (or issuance of a separate permit) for land application of biosolids. Monitoring of the land-applied sludge will be required. The required monitoring frequency, which will be specified in the permit modification, depends on the quantity of sludge disposed of each year. At a minimum, the frequency is as follows.

Amount of Wastewater Sludge (dry tons/year)	Monitoring Frequency
0 to 300	once per year
300 to 1600	once per quarter
1600 to 16 000	once per two months
≥16 000	once per month

OPERATOR CERTIFICATION

Properly trained and certified wastewater treatment plant operators are essential for the protection of public health, preservation of water quality, protection of wastewater treatment facilities, and compliance with environmental regulations. Most states and provinces have the responsibility for ensuring that wastewater treatment personnel are properly certified. Each state and province may have different requirements for the amount of education, training, and experience required to take a certification examination. They may also have different requirements on the level of certification required for a particular plant, and for the number of operators required for operation of the plant. One should check with the local regulatory official about their certification requirements. Table 12.2 is a list of certification contacts in the

Table 12.2 State and provincial certification contacts (http://www.abccert.org/certcontacts.html).

State or province	Contact	Phone number
Alabama	Mark J. Anderson	(334) 279-3040
Alaska	Kenneth A. Smith	(907) 465-5140
Alberta	Kathy Abramowski	(403) 427-7713
Arizona	William Reed	(602) 771-4638
Arkansas	Janet Gay	(501) 682-0998
Atlantic Canada	Paul Klaamas	(902) 426-4378
California	Jennifer Nitta	(916) 341-5639
Colorado	Sharon Ferdinandsen	(303) 692-3558
Connecticut	Joe Nestico	(860) 424-3755
Delaware	R. Peder Hansen, P. E., P. G.	(302) 739-5731
Florida	Edward James	(850) 245-7500
Georgia	Collen Roehm	(478) 207-1450
Hawaii	Marshall Lum	(808) 832-5478
Idaho	Nancy Boswer	(208) 373-0406
Illinois	Robert Voss	(217) 782-9720
Indiana	Natalie Green	(317) 233-0479
Iowa	Dennis Alt	(515) 725-0275
Kansas	Vickie Wessel	(785) 296-2976
Kentucky	Charles Clark	(502) 564-3410
Louisiana	Stacy Williams	(225) 765-5058
Maine	Leslie Rucker	(207) 287-9031
Manitoba	Mike Van Den Bosch	(204) 945-7015
Maryland	Lee Haskins	(410) 537-3167
Massachusetts	Thomas W. Bienkiewicz	(508) 767-2781
Michigan	Eric Way	(517) 373-4752
Minnesota	Dianne Navratil	(651) 296-9269
Mississippi	Kim Smith	(601) 961-5293
Missouri	Gordon Belcher	(800) 361-4827

(continues)

Table 12.2   Continued

State or province	Contact	Phone number
Montana	Jenny Chambers	(406) 444-2691
Nebraska	Steven Goans	(402) 471-2580
Nevada	Diana Silsby	(775) 678-9438
New Brunswick	Andre Chenard	(506) 453-3849
New Hampshire	George C. Neill	(603) 271-3325
New Jersey	Linda Cantwell	(609) 777-1013
New Mexico	Mike Coffman	(505) 827-0188
New York	Phillip Smith	(518) 402-8092
Newfoundland	Ervin McCurdy	(709) 637-2481
North Carolina	Ted Cashion	(919) 733-0026
North Dakota	Craig Bartholomay	(701) 328-6626
Northwest Territories	James Foster	(867) 873-7999
Nova Scotia	John Eisnor	(902) 424-2282
Ohio	Andrew Barienbrock	(614) 664-2752
Oklahoma	Chris Wisniewski	(405) 702-8100
Ontario	Alex Salewski	(416) 314-9387
Oregon	Steve Desmond	(503) 229-5622
Pennsylvania	Kathy Keyes	(717) 787-2043
Prince Edward Island	Morley Fox	(902) 368-5036
Quebec	Pierre Vincent	(514) 270-7110
Rhode Island	Bill Patenaude	(401) 222-4700
Saskatchewan	Gus Feitzelmayer	(306) 787-6174
South Carolina	Dona Caldwell	(803) 896-4430
South Dakota	Robert Kittay	(605) 773-4208
Tennessee	Sherry Messick	(615) 898-8090
Texas	Juanita Lopez	(512) 239-6165
Utah	Judy Etherington	(801) 538-6062
Vermont	Carole Fowler	(802) 241-2369
Virginia	David Dick	(804) 367-8590
Washington	Tammie McClure	(360) 407-6449
West Virginia	Tim Greene	(304) 558-6986
Wisconsin	Peg O'Donnell	(608) 266-0498
Wyoming	Louise Cordova	(307) 777-7781

United States and the Canadian provinces. This list was current at the time of publication of this book. However, the Association of Boards of Certification can provide up-to-date information on the appropriate person to contact in a state or province. Check the website at http://www.abccert.org/certcontacts.html or contact them at (515) 232-3623. Most states will accept the California State Sacramento Operator Training Manual as a prerequisite for certification. Another good source is the Water Environment Federation's® *WEF/ABC Wastewater Operators' Guide to Preparing for the Certification Examination* (WEF, 2002). For additional information on training opportunities, one should check with the local Water Environment Federation® affiliate.

MONITORING AND REPORTING

The permit issued to the plant will probably require specific monitoring and reporting to the appropriate regulatory agency. If it is an NPDES permit issued by U.S. EPA or a delegated state, the reports will probably be submitted on a Discharge Monitoring Report. These forms will be supplied by the regulatory agency, and in many cases, will be preprinted. They may also be printed out by several software packages, which will also perform all of the calculations. It is important to complete the forms and submit them in a timely manner; otherwise, a late report may result in a noncompliance letter from the regulatory agency. These reports typically require submittal of discharge data both in concentration and mass (with some exceptions). They must also be signed by a responsible party. If there are any questions regarding completion of the reports, be sure to contact the regulatory agency for assistance. Most states have an outreach program to provide such assistance.

REFERENCES

American Public Health Association; American Water Works Association; Water Environment Federation (1998) *Standard Methods for the Examination of Water and Wastewater*, 20th ed.; Washington, D. C.

U.S. Environmental Protection Agency (2002a) Section 402—National Pollutant Discharge Elimination System. Federal Water Pollution Control Act (Clean Water Act).

U.S. Environmental Protection Agency (2002b) Title III—Standards and Enforcement, Sections 301–320. Federal Water Pollution Control Act (Clean Water Act).

U.S. Environmental Protection Agency (2000a) Part 122—EPA Administered Permit Programs: The National Pollutant Discharge Elimination System. *Code of Federal Regulations*, Part 60, Title 40.

U.S. Environmental Protection Agency (2000b) Part 133—Secondary Treatment Regulation. *Code of Federal Regulations*, Part 60, Title 40.

U.S. Environmental Protection Agency (2000c) Part 136—Guidelines Establishing Test Procedures for the Analysis of Pollutants. *Code of Federal Regulations*, Part 60, Title 40.

U.S. Environmental Protection Agency (2000d) Part 503—Standards for the Use or Disposal of Sewage Sludge. *Code of Federal Regulations*, Part 60, Title 40.

Water Environment Federation (2002) WEF/ABC Wastewater Operators' Guide to Preparing for the Certification Examination. Water Environment Federation: Alexandria, Virginia.

Chapter 13
Occupational Health and Safety

*I*NTRODUCTION

Occupational health and safety is becoming increasingly important as a business component in the operation and maintenance of water and wastewater facilities. The application of health and safety principles is critical to the proper operation of an extended aeration package plant (EAPP). The ultimate reason for the implementation of safety efforts is the elimination and control of worker injuries and accidents. Along with the pain that a worker feels as a result of a workplace injury, there are additional consequences associated with a workplace injury. Negatives associated with a workplace injury include workers' compensation costs, need for additional training for the worker replacing the injured employee, increase in administrative tasks, decreased morale of the plant workforce, and other results that interrupt the smoothness of a wastewater plant operation.

The implementation of health and safety control measures will assist in continuing and maintaining an organized flow of the process. An accident is any unplanned event which interrupts the smooth flow of a process. Along with maintaining a good process flow in a facility, the elimination and reduction of workplace accidents will help control costs associated with the operation of an EAPP.

In addition to the reduction of pain and the control of costs associated with health and safety efforts, compliance with the Occupational Safety and Health Administration (OSHA) regulations will be improved. While many publicly operated facilities are not required to comply with OSHA regulations, compliance with the regulations is a sound business practice. The implementation of OSHA regulations serves as a guideline for the control of occupational accidents.

The purpose of this chapter is to review the scope of OSHA regulations, summarize previous wastewater plant accident experience, discuss the health and safety process, and review the OSHA regulations that are most applicable to EAPP operation and maintenance.

OCCUPATIONAL SAFETY AND HEALTH REGULATIONS

FEDERAL OCCUPATIONAL SAFETY AND HEALTH ADMINISTRATION. The Federal Occupational Safety and Health Administration (OSHA) was initiated in 1970 as a result of the Occupational Safety and Health Act of 1970 (Occupational Safety and Health Act, 1970). The Administration is responsible for the promulgation and enforcement of health and safety regulations to protect the health and safety of the workers in the workplace in the United States.

The OSHA regulations are found in Title 29 *Code of Federal Regulations* (CFR) Part 1900 to Part 2400. Title 29 is reserved for the Department of Labor, the federal government department where OSHA resides. General industry regulations are found in 29CFR1910 (Occupational Safety and Health Standards (2003) and the construction industry regulations are found in 29CFR1926 (Safety and Health Regulations for Construction, 2003). The OSHA regulations applicable to EAPPs are primarily located in 29CFR1910, but when construction activities are performed at a plant site, the health and safety regulations that apply are contained in 29CFR1926.

The President of the United States appoints, with the approval of Congress, a head of OSHA; the Assistant Director of the Department of Labor. The Occupational Safety and Health Administration is divided into 10 regions across the United States, and the regions are divided into areas. The regions and areas are managed by a region or area director. Compliance health and safety officers (CHSOs) are responsible for the enforcement of the promulgated OSHA regulations in the workplace.

The CHSOs perform workplace inspections for a variety of reasons. The number of inspections in various categories will change based on the emphasis areas and the management of the administration. These reasons for workplace inspections include imminent danger, programmed or special emphasis inspections, fatality/serious injury investigation, employee complaint, and follow-up inspections.

STATE PLAN OCCUPATIONAL SAFETY AND HEALTH ADMINISTRATION PROGRAMS. Currently, there are 23 states with state plan OSHA Programs. These programs are approved by federal OSHA, and the

---

regulations enforced by the state plan OSHA programs must be at least as stringent as the federal OSHA regulations. The primary difference in federal OSHA and state plan OSHA Programs is that federal OSHA is administered by the federal government and the state plans are administered by the states. In many cases, the state plan OSHA programs have regulations that are the same or very similar to federal OSHA. However, there are several states that have promulgated additional health and safety regulations, which they believe are important to protect workers in their state.

It is important for employers to know if a federal or state plan OSHA has jurisdiction for their workplace. Once that is determined, the employer must determine and understand the specific regulations that apply to the workplace.

PRIVATE AND PUBLIC SECTORS REGULATORY COMPLIANCE. While most wastewater plants are presently operated by a public entity, more wastewater plants are contracting with private companies for the operation and maintenance (O&M) of their facilities. Several plants are privatizing, which means that, in addition to being responsible for the O&M of the plant, the private company actually owns the plant. All private companies are subject to OSHA regulations, including these O&M contractors. Currently, public entities are not subject to OSHA regulations or enforcement by CHSOs. In some states, there are state departments that enforce OSHA regulations within the public sector. Regardless, the implementation of OSHA requirements within a private or public operator is a beneficial business practice for the worker and operator.

*O*CCUPATIONAL ACCIDENTS

WASTEWATER TREATMENT PLANT WORKER ACCIDENT EXPERIENCE. Worker Accident Experience. The Bureau of Labor Statistics (BLS) maintains a database of accidents that occur at industrial, service, and construction work locations. These accident rates are maintained in two primary categories; OSHA recordable incident rate and OSHA lost time incident rate. Both of these are incidents per 200 000 employee hours or, on an approximate basis, incidents per 100 employees.

The BLS maintains the database by standard industrial classifications (SIC), a method to distinguish between different industrial, service, and construction businesses. The SIC code for the wastewater industry is 4950. Table 13.1, Accident Rate Comparison, provides a recent review of accident experience for the wastewater industry.

Occupational Safety and Health Administration Recordkeeping Requirements. The OSHA regulation 29CFR1904 requires employers to maintain information concerning employee accident experience (Recording and Reporting Occupational Injuries and Illnesses, 2001). Employers must

Table 13.1 Accident rate comparison (1993).

Industry	Standard industrial classifications code	Recordable incident rate	Lost time incident rate	
Wastewater	4950	13.7	777777	Ed: 777777 as meant?
			7.2	
Water supply	4940	10.4	4.7	
Private industry	—	8.5	3.8	
Construction	15 to 17	12.2	5.5	
Manufacturing	20 to 39	12.1	5.3	

determine if employee accidents are classified as OSHA recordable, lost time, or restricted duty. Also, the employer must distinguish between an occupational injury and an illness. The OSHA regulations 29CFR1904.0 to 1904.46 provide information to the employer to make these decisions.

Once the employer makes the decision on OSHA recordability, recordable accidents must be listed on an OSHA 300 form. This must be completed within five days of the occurrence of the accident. The OSHA 300 form is maintained for a calendar year and the columns are totaled at the end of the calendar year. The employer must post an OSHA 300-A form summary (information without the employees' names) during the months of February to April for the preceding year's recordable injuries and illnesses. The posting should be in the workplace in an area where all employees go during their workshift (i.e., breakroom, timeclock, or message board).

WORKERS' COMPENSATION. Overview. Workers' compensation was put in place to provide monetary compensation to workers who are injured in the workplace. Also, if a worker becomes ill with an occupational disease related to a workplace exposure, then this could be compensatory under workers' compensation laws.

Workers' compensation payments are divided into two general categories; medical and indemnity. Medical payments are to cover medical costs to an injured or ill worker. This would include items such as doctors' fees, prescribed medications, hospital fees, x-rays, and treatments associated with a work-related injury or illness. Indemnity payments are to pay a portion of the employee's salary while he/she is out of work because of a workplace injury or illness. This payment is approximately 66% of an employee's salary and the waiting period is typically five working days away from work, although this differs from state to state. Also, indemnity payments are made for loss of a body part or use of a body part because of a workplace injury. The amount of indemnity compensation in this case is set in the workers' compensation law for that specific state.

As previously alluded to, workers' compensation laws are set by each state. For this reason, it is important that employers understand their state's workers' compensation laws.

There are a variety of workers' compensation insurance plans available in today's market. Insurance companies typically administer an employer's workers' compensation insurance if it is a traditional plan or a self-insured type plan. Again, as with the workers' compensation laws, an employer should have an understanding of the workers' compensation insurance program, as it is typically an important portion of a company's financial picture.

Costs Associated with Workplace Accidents. If a workplace injury or illness occurs, the workers' compensation medical and indemnity payments will be available to the worker, assuming that it is a valid claim. Depending on the incident, these payments can be a significant amount of money. In addition to the workers' compensation costs, there are many hidden costs associated with workplace injuries and illnesses. These hidden costs are typically estimated at four to eight times the workers' compensation costs. These hidden costs consist of the following:

- Training another employee for the job,
- Retraining the injured worker,
- Administrative costs of processing claim,
- Time to conduct incident investigation,
- Stoppage of work, and
- Follow-up efforts with the employee and physician for return to work.

EXPOSURE IDENTIFICATION AND CONTROL MEASURES

In traditional occupational safety and industrial hygiene, the control of accidents, injuries, and illnesses in the workplace is based on the following process: Identify the potential accident or illness hazard, evaluate the potential accident or illness hazard, and implement control measures based on the evaluation portion of the process.

IDENTIFICATION OF POTENTIAL ACCIDENT HAZARDS. The initial effort in the control of accident hazards is the identification of potential and actual accident and injury hazards. There are several ways in which accident hazards are identified. The most critical method is reviewing accident investigations to determine where accidents have previously occurred. These should take priority, as these are situations where workers have already been injured.

Another method to identify potential accident hazards is by conducting a site survey or inspection. In performing this task, a qualified individual will observe equipment and processes to identify sources of accident potential. Table 13.2 indicates the observation area and the potential type of accident that could occur.

Table 13.2   Potential accident hazard identification.

Observation area	Accident type
Guardrail/handrail	Fall from height
Housekeeping	Slips/falls, same level
Machine guarding	Caught in machine
Electrical	Shock/electrocution
Chemicals	Inhalation, skin contact
Mechanical handling	Struck by/against

The third method of identification of accident hazards is to observe workers performing various work tasks. During these observations, efforts should be directed to identify work tasks or subtasks that have the potential for accident or injury. During this observation process, if standard operating procedures (SOPs) are available, they can be used to determine potential hazards.

EVALUATION OF HAZARD. Once potential accident hazards are identified, efforts should then be directed to evaluation of the potential hazards. During the evaluation process, several resources can be used to evaluate the potential of accident or injury from the identified hazards. Industry and OSHA standards and health and safety guidance documents can all be researched to further evaluate the potential for an accident and the lack of control measures. Also, safety professionals and industry associations can be contacted as needed to evaluate the potential hazard.

If numerous hazards are identified, they should be prioritized related to the potential frequency and severity of an accident(s) occurring. This is critical for the next step of the process, control, whereby hazards with the greatest potential for accident frequency and severity can be controlled quicker than lower potential hazards.

Also, during this phase of the process, some exposures will need to be evaluated by sampling. This is especially true when potential chemical exposures or hazards are identified.

CONTROL OF ACCIDENT HAZARD. Once hazards have been identified and evaluated, efforts should be directed to control measures. It is important that the hazards be evaluated and prioritized if needed before control measures are implemented. The development of control measures will typically take individuals from various disciplines, including safety, process, operations, engineering, and maintenance. The use of personnel from various disciplines will help insure that the control measure is practical and effective for all personnel at a facility.

Elimination of Hazard. Elimination of the hazard is the best control measure and should be the objective of all personnel involved with the development of control measures. Examples of this strategy include substitution of a low toxicity chemical for a chemical higher in toxicity, substitution of a nonflammable chemical for a flammable chemical, or replacement of equipment without the identified hazard. It can be seen that, while this is the most desirable control measure, it is very difficult to implement a practical and feasible elimination of a hazard.

Engineering Controls. Engineering controls are highly desirable control measures and there are many opportunities within the workplace to implement these control measures. Examples of engineering controls include installation of guardrails or handrails, local exhaust ventilation for airborne chemicals, installation of mechanical handling aids, installation of machine guards, and installation of ground fault circuit interrupters.

Administrative Controls and Training. Administrative controls are important control measures, but are rarely successful without elimination or engineering control measures. They should be used to supplement the engineering control measures. These controls are the development of various health and safety programs, along with SOPs. Examples of health and safety programs include housekeeping, confined space entry, and hot work. Once developed, training should be implemented for each worker who is potentially exposed to a hazard that the program strives to control. All training should be performed on a periodic basis to verify that workers have an understanding of the potential hazards and control measures.

At this point of the control process, there is a need to develop enforcement methods of the engineering and administrative controls. The education portion is extremely important, but workers should be required to abide by the administrative controls that have been developed.

Personal Protective Equipment. The least desirable of all control options is personal protective equipment (PPE). The use of PPE is still necessary in the workplace to be used in conjunction with engineering or administrative control measures. Also, there are hazards where engineering controls are not practical or feasible and control measures must be a combination of administrative and PPE control measures. As with the administrative control measures, a method of enforcement must be implemented.

It is highly desirable to use PPE as a temporary control measure when engineering controls are being implemented.

HEALTH AND SAFETY PROGRAM ELEMENTS

The base of quality accident prevention and OSHA compliance is a written health and safety program. The written program provides guidance and direction for all of the accident prevention and OSHA compliance efforts at an EAPP. The complexity of the written health and safety program should be determined by the size and/or complexity of the facility and the potential hazards that the workers are exposed to during the performance of their work tasks. Regardless of how large or small a facility is, there is a need for some level of a health and safety program.

STATEMENT OF COMPANY POLICY. The initial portion of a written health and safety program should indicate a statement of company/authority policy. This would be directed to health and safety and be a commitment as to the level of safety efforts and the responsibilities of various personnel within the organization. In addition to this statement being in writing, it is critical that this statement be followed at all times by personnel within the organization.

RESPONSIBILITIES. Employer. The employer has a duty to provide a safe and healthy workplace for each employee. Section 5(a)(1) of the Occupational Safety and Health Act (P.L. 91-596) is referred to as OSHA's general duty clause. It states, "Each employer shall furnish to each of his employees employment and a place of employment which is free from recognized hazards that are causing or are likely to cause death or serious physical harm to his or her employees (Occupational Safety and Health Act, 1970)."

Management. Management has the responsibility to determine and verify that the health and safety programs and control measures are being followed by all employees. Management should conduct or coordinate health and safety training efforts to educate employees concerning hazards and control measures. Once the training is complete, management's duty is to enforce the use of control measures to reduce and eliminate hazards to employees.

Employees. Employees have the responsibility to follow safety requirements and procedures. Also, employees should assist management in identifying potential accident hazards at the EAPPs. The purpose of these responsibilities is for the employee to reduce the potential of an accident to the employee or to a fellow worker.

REGULATORY REQUIREMENTS. Many OSHA regulations require a written program and/or procedures to comply with the regulation. These include 29CFR1910.1200—Hazard Communication (1983), 29CFR1910.146—Permit Required Confined Space Entry (1993), 29CFR1910.147—Control of Hazardous Energy (Lockout/Tagout) (1989), and 29CFR1910.38—Medical and Emergency Services (1980). Operators of EAPPs should be familiar with the OSHA regulations in the Identified Accident Hazards, Controls, and Regulatory Requirements section of this chapter to determine what type of written program is required to comply with the regulation.

RECORDKEEPING. Proper recordkeeping is critical to a quality health and safety program. Records must be kept on a variety of activities from safety inspection reports to accident reports to various regulatory requirements. These are important to management to verify that key components of a safety program are implemented and critical to provide documentation to OSHA in the event of an OSHA regulatory inspection.

SITE INSPECTIONS AND AUDITS. To verify that proper health and safety controls are in place, periodic site inspections or audits should be conducted. In addition to determining that adequate controls are in place to protect the worker, these audits also can be used to determine compliance with OSHA regulations. The audits can range from comprehensive one completed by a safety professional to a basic audit conducted by the operator of the plant. Regardless, these audits are key elements in a health and safety program. Table 13.3 provides an example checklist that can be used to conduct a periodic check of the safety controls in place at an EAPP.

Table 13.3   Health & safety checklist.

Plant: _____   Date: _____
Operator: _____

Safety control	Yes	No	NA	If no, correction date	Comments
Lockout devices— available and used					
Confined space entry equipment—available and used					
PPE—Hand protection					
PPE—Eye protection					
PPE—Safety shoes/boots					
PPE—Hard hats					
Guardians in place					
Handrails in place					
Housekeeping (trip and fall hazards)					
Signage—PRCS					
Signage—No smoking					
MSDS available					
Pump guards in place					
Other guards in place					
Fire extinguishers					

PPE = personal protective equipment
PRCS = permit required confined space
MSDS = material safety data sheets

IDENTIFIED ACCIDENT HAZARDS, CONTROLS, AND REGULATORY REQUIREMENTS

HAZARD COMMUNICATION. Hazard communication is the process whereby chemical manufacturers, distributors, and users communicate chemical hazard information and control measures downstream. The OSHA Hazard Communication standard, 29CFR1910.1200, went into effect in 1986. The standard has requirements to meet the intent of the standard, pass chemical information, and control measures from the chemical manufacturer to the end user. This includes material safety data sheets (MSDS), container labeling, and information and training.

Material Safety Data Sheets. Material safety data sheets are documents that contain information about the specific chemical or chemical product. In 29CFR1910.1200, OSHA has specified the type of information that is required on the MSDS, and it includes the following information (Hazard Communication, 1983):

(1) Identity used on the label.
(2) Chemical and common name(s).
(3) If the material is a mixture, the chemical and common names of
    (a) Hazardous Ingredients >1% of composition.
    (b) Carcinogens >0.1% of composition.
(4) Physical and chemical characteristics (vapor pressure, flash point).
(5) Physical hazards (fire, explosion, reactivity).
(6) Health hazards.
    (a) Signs and symptoms of exposure.
    (b) Medical conditions aggravated by exposure.
(7) Primary route(s) of entry.
(8) Exposure limits.
    (a) OSHA Permissible Exposure Limit.
    (b) American Conference of Governmental Industrial Hygienists (ACGIH) Threshold Limit Value.
    (c) Any other exposure limit that is used.
        (i) If hazardous chemical is listed as a carcinogen by
(9) National Toxicology Program (NTP).
(10) International Agency for Research on Cancer (IARC).
(11) OSHA.
(12) Any generally applicable precautions for safe handling, including
    (a) Appropriate hygienic practices.
    (b) Protective measures during repair and maintenance.
    (c) Procedures for clean-up of spills and leaks.

---

(13) Any generally applicable control measures, including
    (a) Engineering controls.
    (b) Work practices.
    (c) Personal protective equipment.
(14) Emergency and first aid procedures.
(15) Date of preparation or revision.
(16) Name, address, and telephone number of party responsible for preparing the MSDS.

The Occupational Safety and Health Administration does not have a standard form that is required for the MSDS, but the information listed above must be included on the form that is used. The MSDS must be available on any material that is or has the potential to create a chemical or physical hazard to personnel under normal use or in emergency situations. Chemical manufacturers and distributors are required to provide this information downstream to purchasers of the material. The MSDS must be readily available for workers to use as needed.

Labeling and Hazard Warnings. The hazard communication standard requires that all containers of materials that create a potential chemical or physical hazard have proper labeling. Typically, containers of chemicals received from a vendor have the proper labels in place. Regardless, it is the responsibility of the user of the chemical to verify that the label is in place. Also, except with immediate use containers, if hazardous chemicals are transferred to another container on site, then that container must be labeled. The hazard communication standard requires that the label contain the identity of the chemical and the appropriate hazard warnings.

Information and Training. The hazard communication standard requires that the employer have an information and training program for employees that are exposed or potentially exposed to hazardous chemicals. The information and training program should contain the following elements:

(1) Information.
    (a) Any operations in the employee's work area where hazardous chemicals are present.
    (b) The location and availability of the written hazard communication program, including:
        (i) List of hazardous chemicals.
        (ii) MSDSs.
(2) Training.
    (a) Methods and observations used to detect the presence or release of hazardous chemicals in the workplace.
    (b) Physical and health hazards of the chemicals in the work area.

(c) The measures that employees can take to protect themselves, including
  (i) Work practices.
  (ii) Emergency procedures.
  (iii) Personal protective equipment.
(d) Details of the hazard communication program, including
  (i) Explanation of the labeling system.
  (ii) Explanation of MSDSs.

PERMIT-REQUIRED CONFINED SPACE ENTRY. Definitions. The OSHA Permit Required Confined Space Entry Standard (29CFR1910.146) (1993) regulates entry into permit-required confined spaces. There are critical definitions in the standard relating to definitions of a permit required confined space and roles that individuals perform during confined space entry. The most important item to initially determine is if the space is a permit-required confined space. First, one must determine if the space is a confined space by the definition below. The space must meet all three of the criteria listed below. The most important of the criteria is having limited means of entry and exit. This is typically a manway, fixed vertical ladder, or portable ladder. Standard stairs or a standard doorway do not meet the definitions of limited means of entry and exit. A space cannot be a permit-required confined space if it is not a confined space.

Confined space. A space or work area not designed or intended for normal human occupancy, having limited means of entry and exit; and large enough and so configured that an employee can bodily enter and perform assigned work.

Once the space is determined to be a confined space, one must determine if it is a permit-required confined space. A confined space must meet only one or more of the following four areas to be a permit-required confined space. The definition of hazardous atmosphere is also important, and the confined space must only have the potential of one of the hazardous atmospheres to be considered a permit-required confined space.

Permit-required confined space (permit space). A confined space that has one or more of the following characteristics:

(A) Contains or has the potential to contain a hazardous atmosphere.
(B) Contains a material that has the potential for engulfing an entrant.
(C) Has an internal configuration such that an entrant could be trapped or asphyxiated by inwardly converging walls or by a floor which slopes downward and tapers to a smaller cross section.
(D) Contains any other recognized serious safety or health hazard.

Hazardous atmosphere. An atmosphere with the following conditions:

(A) Greater than 10% of the lower explosive level (LEL).
(B) Airborne combustible dust greater than LEL (dust obscures vision at a distance of 1.524 m [5 ft] or less).

(C) Atmospheric oxygen concentration is less than 19.5% or greater than 23.5%.

(D) Atmospheric concentration of a substance, which could result in employee exposure greater than the permissible exposure limit.

(E) Any atmospheric condition that is immediately dangerous to life and health.

In EAPPs, the primary items that make a confined space a permit confined space is the potential for a hazardous atmosphere or the potential for engulfment. Possible permit required confined spaces at EAPP operations include tanks, basins, sewers, and wet and dry wells. These are potential permit-required confined spaces, and the definitions must be studied to verify this is the case.

The following are definitions of roles of individuals during permit-required confined space entry. It is important to understand that the minimum number of people to conduct permit required confined space entry is two. If the confined space team consists of two individuals, one individual must assume two of the roles listed below.

Attendant. The individual stationed outside of the permit space who monitors the authorized entrants.

Authorized entrant. An individual who enters the permit space.

Entry supervisor. The individual responsible for determining acceptable entry conditions exist, for authorizing entry, overseeing entry operations, and for terminating entry as needed. The entry supervisor may also serve as the attendant or authorized entrant.

Identification of Permit-Required Confined Spaces. The identification of permit required confined spaces at EAPPs is an important portion of a health and safety program. Once the permit-required confined spaces are identified, they should be listed in the site confined space program, and labeled "Danger— Permit Required Confined Space, Do Not Enter."

Entering Permit-Required Confined Spaces. A team of two or three qualified individuals is necessary to conduct confined space entries. Site-specific procedures must be set up at each EAPP to protect the health and safety of the workers performing these tasks. The following are general procedures that should be considered before and during entry tasks:

(1) Implement measures necessary to prevent unauthorized entry, for example, locking of entry locations.

(2) Acceptable entry conditions should be determined based on the characteristics of the permit space.

(3) The permit space should be isolated and completely protected against the release of energy or material. This can be accomplished by blanking, blinding, double block and bleed system, lockout of all energy

sources, blocking or disconnecting mechanical linkages, and misaligning or removing sections of lines, pipes, or ducts.

(4) Barriers should be provided to protect entrants from external hazards.

(5) The proper equipment and PPE should be adequately maintained, calibrated before and after entry, and employees who use the equipment should be adequately trained.

(6) The permit space should be monitored remotely before entry into permit space.

(7) At least one attendant is in contact with the entrant through the entire entry process.

(8) Rescue and emergency services should be determined before entry. Nonentry rescue (full body harness with retrieval system) should be provided if a hazard is not generated by the rescue equipment. If nonentry rescue is not possible, then rescue procedures shall be determined. All rescue and emergency procedures should be practiced at least annually in emergency drills.

(9) Entry permits should be prepared and used during each permit space entry. The permits should be written and kept on file for a minimum of one year. Figure 13.1 shows an example entry permit, the most important health and safety guidance mechanism available to the workers.

Training. Workers who perform permit-required confined space entry tasks should be adequately trained for their role (entrant, attendant, or entry supervisor) in the entry operation. This would include procedures, PPE, physical and/or chemical hazards, air monitoring, and rescue.

WALKING/WORKING SURFACES AND FALLS. Fall Prevention. Efforts should be directed to the control of the two types of falls; on the same level and from heights. Fall prevention efforts for falls on the same level involve the floor/ground surface. This includes good housekeeping and having level walking surfaces. The walking surfaces should be free from changes in level (except the use of stairs or steps) and trip hazards (electrical conduit, hoses) on the walking surface.

Falls from height are controlled by protecting walking surfaces at heights by a physical fall prevention method. This is typically a standard guardrail consisting of an upper rail and midrail. A toeboard is required if there is a potential of a worker below the area and materials could be kicked off or fall off and strike the worker. A handrail(s) is needed for stairs that have four or more steps. Guardrails or handrails should be in place when there is a potential fall from height or into water. See 29CFR Subpart D—Walking and Working Surfaces (1970).

Fall Protection. Fall protection is related to protection of workers from falls from heights when there is not a physical fall prevention device in place. Fall protection would involve the use of a full body harness and lanyards, full body harness with a third rail system on fixed vertical ladder, or other methods to protect an employee from fall from height.

_____ **PLANT**

1. GENERAL ENTRY INFORMATION
 	
Permit Space
To be Entered:_____
 	
_____
 	
Date/Time Issued: _____
 	
Permit Duration: _____
 	
Purpose of Entry: _____
 	
2. Confined Space Participants List (print names):
 	
Entrants(s)   1. _____
            2. _____
            3. _____
Attendant(s) 1. _____
            2. _____
            3. _____
Supervisor(s) 1. _____
            2. _____
            3. _____
	3. Communication Procedures
	To be used by Attendants and Entrants:
	
	_____
	
	4. Permit Space Hazards (Actual or Potential)
	
	Atmospheric Hazards:
	    Oxygen deficiency (<19.5%)
	    Oxygen enrichment (>23.5%)
	    Flammable gases or vapors (>10% of LEL)
	    Toxic gases or vapors (>PEL/TLV)
	    Airborne combustible dusts (> or = LEL)
	Physical Hazards:
	    Engulfment (flooding/cave-ins)
	    Mechanical hazards/moving parts
	    Exposure to live electrical parts
	    Materials harmful to skin
	    Specify _____
	
	5. Emergency and Rescue Services
	
	Name of Service: _____
	Emergency Phone Number: _____
	Non-Emergency: _____

6. Required Safety Equipment

	Yes No		Yes No
Multi-gas Detector Full Body Harness and Lifeline for Entrants Rescue Retrieval Winch/Fall Arrest Equipment Portable Ladder Two-Way Radios Respirators: _____ Explosion-proof lighting and tools		Hardhat Steel-toe boots Safety Glasses Hearing Protection Gloves:_____ Protective Clothing:	

7. Site Control and Monitoring Procedure

A. Pre-Entry
  1. Atmospheric Checks (pre-ventilation):
     Time: _____ (Record results in table)

  2. Source Isolation (No Entry):   Yes No
     Pumps or Lines Blinded
     Lines double-blocked/bled off
     Electrical/Mechanical Lock & Tag
  3. Ventilation:
     Mechanical (Forced Air)
     Natural Ventilation

B. Requirements Completed Before Entry: Time
   Lockout/Tagout of required equipment _____
   Purging/Flushing if necessary _____
   Secure Area (Post permit and signs) _____
   Pedestrian & Vehicle Guards/Barriers _____

C. Atmospheric Checks After Isolation and Ventilation:
   Time: _____ (Record results in table)

D. Sampling Equipment:

   Equipment: _____
   Equipment Serial #: _____
   Calibration Date: _____
                                 Yes No
E. Communications check: verified? Y

F. Other Equipment/Permits
                                 Yes No
   Hot Work/Welding Permit
   Fire Extinguisher

cs permit   all sections must be filled out

Figure 13.1   Confined space entry permit.

LOCKOUT/TAGOUT—CONTROL OF HAZARDOUS ENERGY. Identification of Energy Sources. Lockout/tagout does not involve only electrical energy sources, but in OSHA 29CFR1910.147 (1989), an energy source is any source of electrical, mechanical, hydraulic, pneumatic, chemical, thermal, or other energy. The initial effort in an effective lockout/tagout program is to identify all of the energy sources and the associated controls of those sources.

Lockout/Tagout Procedures. After the identification of energy sources there is a need to develop procedures to lockout the energy sources before maintenance

8.   ATMOSPHERE TEST AND MONITORING RECORD

	Hazards (record all results)						
		Toxic and/or asphyxiating gases					
	Oxygen	LEL	H₂S	CO			
Time	**Acceptable Entry Conditions**						
Hr:min	19.5% to 23.5%	<10% LEL	<10 ppm	<25 ppm			Tester Initials
Pre-Ventilation							
Post-Ventilation							
Entry							
00:15							
00:30							
00:45							
01:00							
02:00							
03:00							
04:00							
05:00							
06:00							
07:00							
08:00							
Acceptable Entry Condition		<10% LEL	<10 ppm	<25 ppm			

9.   CONFINED SPACE SUPERVISOR ENTRY AUTHORIZATION

Are All Pre-entry Conditions (Sections 1 to 8 satisfied)?   Yes   No

Signature _____   Time/Date: _____

10.   PERMIT CANCELLATION

Job Completed Normally   Job Aborted   Reason:

Signature _____   Time/Date: _____

Cancelled permit must be returned to Safety Coordinator for site filing

All sections must be filled out

Figure 13.1   Continued.

tasks are initiated on a piece of equipment. To protect maintenance personnel from injury, the lockout/tagout procedures should be strictly adhered to.

EMERGENCIES. Emergency Action Plan. Each EAPP needs to have a site-specific emergency action plan (EAP). The EAP should address any potential emergency that could occur at the EAPP to include fire, chemical release, hurricane, tornado, bomb threat, injuries, or any other potential disaster associated with that facility. The EAP should address what actions all personnel will take in an emergency and who has responsibilities in these areas.

Emergency Response. In 29CFR1910.120—Hazardous Waste Operations and Emergency Response (1986), OSHA defines emergency response as a response effort by employees from outside the immediate release area or by other designated responders (i.e., mutual-aid groups and local fire departments) to an occurrence, which results, or is likely to result, in an uncontrolled release of a hazardous substance. Responses to incidental releases of hazardous substances, where the substance can be absorbed, neutralized, or otherwise controlled at the time of release by employees in the immediate release area, or by maintenance personnel, are not considered to be emergency responses within the scope of this standard. Responses to releases of hazardous substances where there is no potential safety or health hazard (i.e., fire, explosion, or chemical exposure) are not considered to be emergency responses.

There are significant training requirements for emergency responders and a large amount of equipment is needed for these efforts. Typically, EAPP operators are not trained in emergency response and do not have the equipment to perform this task. If an EAPP has gaseous chlorine or sulfur dioxide on site there is a need to have a qualified emergency responder available. This need is best filled by a local fire department, hazardous materials (HazMat) team, or contractor with these qualifications and capabilities.

GASEOUS CHLORINE AND SULFUR DIOXIDE. Gaseous chlorine and sulfur dioxide are the two most significant hazardous chemicals used at EAPP, with gaseous chlorine being the most commonly used method of disinfection. These two gases are not only hazardous to on-site operators, but each is a threat to the environment and the public.

Control Measures. While it is, at times, a long-term process, the best control technology for gaseous chlorine and sulfur dioxide is elimination or substitution. The elimination of either or both of these materials with chemical substitution or process modification is by far the best control method. The most common chemical substitutions at EAPPs for these hazardous chemicals are sodium hypochlorite for disinfection and sodium metabisulfite for dechlorination. While each of these chemicals is hazardous, the degree of hazard is much less with the use of these materials. Sodium hypochlorite and sodium metabisulfite are typically used in a liquid form, and potential airborne levels of these materials are minimal when compared to gaseous chlorine and sulfur dioxide. When sodium hypochlorite and sodium metabisulfite are used, consideration must be given to the prevention and control of spills. Handling procedures to be used with these materials would be proper personal protective equipment to reduce the potential of skin contact with these chemicals.

There are other process modifications to eliminate the use of gaseous chlorine and sulfur dioxide, but these are not common to EAPPs. These other process modifications include the use of ultraviolet and ozone disinfection techniques.

If gaseous chlorine and/or sulfur dioxide are used at the EAPP, then appropriate control measures should be in place. The integrity of piping, feed lines, etc., associated with the process must be continually checked for leaks or potential leaks. A gas-monitoring system specific to these materials with alarms should be in place to warn operators of a leak. Vacuum gauges should be functional and a reduction in vacuum pressure should automatically shut down the feed system.

Handling Procedures. Chlorine and sulfur dioxide handling procedures are critical to the control of potential releases of these materials. Standard operating procedures should be in place from the receipt of the materials from the vendor to changeout of the cylinders to storage and loading of the used cylinders.

The use or availability of escape-only respiratory protection devices during cylinder changeout is a good standard practice. The potential for a release is greatest during cylinder changeout, and the use of an escape respiratory protection device will reduce to potential of a severe respiratory injury in the case of a release. The use of a self-contained breathing apparatus can also be used during cylinder changeout and will provide for greater protection to the operator.

The safe handling of the cylinders should be considered to reduce the potential of a leak or rupture during this activity. The securing of cylinders during material handling and when the cylinder is stored is another important control to reduce the potential of a cylinder rupture.

Emergencies. Emergencies during the handling and use of gaseous chlorine and sulfur dioxide can be significant and potentially catastrophic. It is typically beyond the capabilities of EAPP operators to respond to emergencies of these gaseous materials. As reviewed earlier in this chapter, a qualified HazMat team should be identified to assist the plant in a chemical release emergency.

ELECTRICAL HAZARDS AND SAFETY. The primary potential injuries associated with electrical hazards at an EAPP are electrical shock and electrocution. All electrical outlets near wet or damp areas should be equipped with ground fault circuit interrupters, and all portable electric tools should be properly grounded or be double insulated.

High-voltage electrical panels (greater than 600 volts) should be labeled "High Voltage", and only qualified electricians with the proper equipment should work on these systems.

MACHINE GUARDING. While there are several types of equipment that need adequate guarding in an EAPP, the most prevalent are various pumps. Typically, there is a need for shaft and coupling guards on pumps. These should fully enclose the shaft or coupling and only be removed if the pump is properly locked out. If lubrication is necessary, a grease fitting should be installed to an area outside the guard so that lubrication can take place without the removal of the guard.

BIOHAZARDS. A newly identified area of potential exposure and concern during the operation of an EAPP is biohazards. At this time there are no regulations, with the exception of 29CFR1910.1030 Bloodborne Pathogens (1991), that deal with worker protection from biohazards. While there are varying opinions, most research has indicated a limited potential exposure to wastewater treatment operators from bloodborne pathogens. Regardless, efforts should still be directed to control this remote exposure potential and to control potential exposures to workers from other biohazards.

Currently, the best control measure at this point is to eliminate skin contact and ingestion of materials potentially containing biohazards. This is completed by the use of hand protection and additional protective clothing as needed, depending on the task that is being completed. Once the task is completed, good personal hygiene practices are critical to remove any materials on the skin to eliminate the potential of ingestion of the materials during eating, drinking, or smoking.

Summary

Occupational safety and health, while sometimes complicated, is extremely important to the operation and maintenance of an EAPP. This chapter has provided an overview of health and safety, and knowledge of health and safety regulations and requirements is critical to the operator of these systems. Safety and health improvements will reduce the potential of an accident or injury, improve OSHA compliance, and reduce other liabilities associated with the O&M of an EAPP.

References

Bloodborne Pathogens (1991) *Code of Federal Regulations*, Part 1910.130, Title 29, December.

Control of Hazardous Energy (Lockout/Tagout) (1989) *Code of Federal Regulations*, Part 1910.147, Title 29, September.

Hazard Communication (1983) *Code of Federal Regulations*, Part 1910.1200, Title 29, November.

Hazardous Waste Operations and Emergency Response (1986) *Code of Federal Regulations*, Part 1910.120(q), Title 29, May.

Medical and Emergency Services (1980) *Code of Federal Regulations*, Part 1910.138, Title 29, September.

Occupational Safety and Health Act (1970) *U.S. Code*, Section 5(a)(1), Title 29, PL91-596.

Occupational Safety and Health Standards (2003) *Code of Federal Regulations*, Part 1910, Title 29.

Permit Required Confined Space Entry (1993) *Code of Federal Regulations*, Part 1910.146, Title 29, January.

Recording and Reporting Occupational Injuries and Illnesses (2001) *Code of Federal Regulations*, Part 1904, Title 29, January.

Walking and Working Surfaces (1970) *Code of Federal Regulations*, Subpart D, Title 29.

Suggested Readings

Water Environment Federation (1991) *Biological Hazards at Wastewater Treatment Facilities*; Water Environment Federation: Alexandria, Virginia.

Water Environment Federation (1998) *Confined Space Entry*, 2nd ed.; Water Environment Federation: Alexandria, Virginia.

Water Environment Federation (1998) *Operation of Municipal Wastewater Treatment Plants*, 5th ed. (MOP No. 11); Water Environment Federation: Alexandria, Virginia.

Index

A

Accident hazard control, 157
Accident investigation, 156
Accident rate comparison, 155
Accidents, 154
   cost, 156
Activated sludge
   bacteria, 49
   conversion, 47
   flocculation, 47
   microscopic examination, 60
   theory, 46
   transfer, 47
Activated sludge formation, 47
Activated sludge growth curve, 52
Activated sludge process control, 53
Activated sludge system, 32
Additives, 14
Administrative controls and training, 158
Advanced unit processes, 117
Aerating and mixing capability, 35
Aeration chamber, 31, 116
Aeration cycle, 42
Aeration period, 56
Aeration tank, 34
   configuration, 35
   oxygen level, 54
   sizing, 35
Aerator blade types, 41

Air lift pump, 17
Alkalinity, 68
Alum, 126
Aluminum, 135
Ammonia, 127
Ammonia-nitrogen concentration, 67
Amoeboids, 61
Anodes, 8
Association of Boards of Certification, 148

B

Backwashing facilities, 121
Bacteria
   declining-growth phase, 53
   dispersed growth, 53
   log-growth phase, 53
Bacterial growth, 51
Bacterial population dynamics, 50
Bar screen, 23, 116
Biochemical oxygen demand, 13
Biohazards, 169
Biosolids, 90, 146
Blower-motor arrangement, 38
Blowers, 39
Boat waste, 16
Building maintenance, 132
Bulking, 64

C

Calcium hypochlorite, 96
Cell lysis, 44
Cell residence time, 58
Cell yield, 68
Channeling, 83
Chemical attack, 129
Chemical control measures, 168
Chemical emergencies, 169
Chemical handling procedures, 169
Chemical oxygen demand, 14
Chlorination, 93
Chlorinators
   gas, 95
   liquid, 96
   solid tablet or pellet, 97
Chlorine contact tank, 116
Circular clarifier, 77
Clarifier design criteria, 79
Clarifier operation, 82
Clarifier, 73, 116
   troubleshooting, 84
   water temperature, 84
Clearwell, 122
Climate conditions, 129, 136
Clumping sludge, 85
Coal tar epoxy, 132
Color test, 111
Commercial discharges, 145
Comminutor, 25, 116
Concrete, 134
Confined-space entry, 164
Confined-space entry permit, 165
Constant pump head, 12
Control of hazardous energy, 165
Corrosion, 6
Corrosion control, 129

D

Dechlorination, 100
Denitrification, 68, 85, 128
Dewatering, 8

Differential aeration cells, 131
Diffused air, 35
Dilution test, 107
Discharge baffle, 82
Discharge Monitoring Report, 149
Disinfection, 93
Dissolved oxygen, 111
Disinfection, control systems, 99
Dissolved oxygen ampoules, 114
Dissolved oxygen level, 54
Dissolved oxygen meter, 113
Dosing, 118
Downfeed pipes, 36
Drop titration colorimetric test, 114

E

Eductor, 17
Effluent limitations, 144
Effluent sampling, 145
Electrical hazards, 169
Electrical protection, 28
Electrochemical corrosion, 130
Electrolytic generation, 101
Electrolytic solutions, 130
Emergencies, 167
Emergency action plan, 167
Emergency response, 167
Employee responsibilities, safety, 159
Endogenous respiration, 53
Engineering controls, 158
Entering permit-required confined
    spaces, 164
Equalization tank overflow, permit
    requirements, 19
Equalization tank, 116
Exposure identification, 156

F

Fall prevention, 165
Fall protection, 165
Federal Reference Method 24, 140

S

Safety, 138, 151
    employer responsibilities, 159
    management responsibilities, 159
    recordkeeping, 160
    regulations, 153
Safety audits, 160
Safety training, 159
Scouring velocity, 17
Screening media, 119
Scum removal, 83
Secondary biological treatment, 46
Settleability, 106
Settleometer test, 107
Settling curve, 108
Seven-day moving average, 57
Site inspections, 160
Site location factors, 5
Site, groundwater, 7
Skimmer line, 45
Skimming and scum removal, 83
Sludge age, 58, 85, 107
Sludge blanket depth, 74
Sludge disposal, 90
Sludge drawoff, 81
Sludge inventory, 57
Sludge volume index, 59
Sludge wasting, 34, 87
Sodium hypochlorite, 96, 168
Sodium metabisulfite, 168
Solids disposal, 146
Solids loading rate, 80
Solids management program, 87
Solids overflow, 85
Solids retention time, 59
Solids washout, 44
Solubility of oxygen, 113
Specific oxygen uptake rate, 47, 60
Staining methods, 65
Stalked ciliates, 62
State safety regulations, 153
Statement of Company Policy, 159
Steel, 134
Submergence, 17
Sulfur dioxide, 100, 168
Supernating, 89

Surface aeration, 41
Surface settling rate, 79

T

Tertiary treatment, 126
Time clocks, 41
Total suspended solids, 14
Toxicity effects on treatment
    plants, 15
Toxicity, 145
Trash trap, 26
Types of clarifiers, 75
Typical operator visit, 115

U

Ultraviolet disinfection, 101, 115
Uncontrolled release, hazardous
    substance, 168

V

V-belts, 40
Volatile organic content, 140
Volatile suspended solids, 60
Volume solids, 139

W

Walking surfaces, 165
Waste activated sludge, 34, 54
Wastewater characteristics, 13, 50
Wastewater temperature, 47
Water density, 84
Weir overflow rate, 79
Winter proofing, 138
Work pot life, 140
Workers' compensation, 155
Worms, 62